This is my math book.

Name: _______________________

CONTENTS

Date : _________

Time : _________

Name : _________

Score : _____/60

Addition Worksheet 0 - 5

1) 2 + 3 = _____

2) 1 + 1 = _____

3) 1 + 3 = _____

4) 2 + 1 = _____

5) 2 + 3 = _____

6) 1 + 3 = _____

7) 1 + 1 = _____

8) 1 + 4 = _____

9) 4 + 1 = _____

10) 1 + 1 = _____

11) 2 + 2 = _____

12) 2 + 1 = _____

13) 3 + 2 = _____

14) 3 + 1 = _____

15) 1 + 4 = _____

16) 1 + 2 = _____

17) 2 + 2 = _____

18) 0 + 4 = _____

19) 0 + 1 = _____

20) 1 + 1 = _____

21) 1 + 4 = _____

22) 0 + 5 = _____

23) 2 + 3 = _____

24) 4 + 0 = _____

25) 0 + 4 = _____

26) 5 + 0 = _____

27) 4 + 1 = _____

28) 1 + 3 = _____

29) 1 + 3 = _____

30) 4 + 1 = _____

31) 2 + 3 = _____

32) 2 + 1 = _____

33) 3 + 1 = _____

34) 4 + 1 = _____

35) 2 + 3 = _____

36) 5 + 1 = _____

37) 1 + 3 = _____

38) 5 + 0 = _____

39) 1 + 4 = _____

40) 1 + 1 = _____

41) 1 + 1 = _____

42) 2 + 1 = _____

43) 2 + 3 = _____

44) 1 + 3 = _____

45) 2 + 1 = _____

46) 3 + 1 = _____

47) 1 + 2 = _____

48) 2 + 2 = _____

49) 4 + 2 = _____

50) 0 + 5 = _____

51) 2 + 0 = _____

52) 2 + 1 = _____

53) 2 + 2 = _____

54) 4 + 0 = _____

55) 3 + 2 = _____

56) 3 + 2 = _____

57) 5 + 0 = _____

58) 2 + 1 = _____

59) 4 + 1 = _____

60) 2 + 3 = _____

Answer Key
Addition Worksheet 0 - 5

1) 2 + 3 = **5**

2) 1 + 1 = **2**

3) 1 + 3 = **4**

4) 2 + 1 = **3**

5) 2 + 3 = **5**

6) 1 + 3 = **4**

7) 1 + 1 = **2**

8) 1 + 4 = **5**

9) 4 + 1 = **5**

10) 1 + 1 = **2**

11) 2 + 2 = **4**

12) 2 + 1 = **3**

13) 3 + 2 = **5**

14) 3 + 1 = **4**

15) 1 + 4 = **5**

16) 1 + 2 = **3**

17) 2 + 2 = **4**

18) 0 + 4 = **4**

19) 0 + 1 = **1**

20) 1 + 1 = **2**

21) 1 + 4 = **5**

22) 0 + 5 = **5**

23) 2 + 3 = **5**

24) 4 + 0 = **4**

25) 0 + 4 = **4**

26) 5 + 0 = **5**

27) 4 + 1 = **5**

28) 1 + 3 = **4**

29) 1 + 3 = **4**

30) 4 + 1 = **5**

31) 2 + 3 = **5**

32) 2 + 1 = **3**

33) 3 + 1 = **4**

34) 4 + 1 = **5**

35) 2 + 3 = **5**

36) 5 + 1 = **6**

37) 1 + 3 = **4**

38) 5 + 0 = **5**

39) 1 + 4 = **5**

40) 1 + 1 = **2**

41) 1 + 1 = **2**

42) 2 + 1 = **3**

43) 2 + 3 = **5**

44) 1 + 3 = **4**

45) 2 + 1 = **3**

46) 3 + 1 = **4**

47) 1 + 2 = **3**

48) 2 + 2 = **4**

49) 4 + 2 = **6**

50) 0 + 5 = **5**

51) 2 + 0 = **2**

52) 2 + 1 = **3**

53) 2 + 2 = **4**

54) 4 + 0 = **4**

55) 3 + 2 = **5**

56) 3 + 2 = **5**

57) 5 + 0 = **5**

58) 2 + 1 = **3**

59) 4 + 1 = **5**

60) 2 + 3 = **5**

Date : _________

Time : ________

Name : __________

Score : _____/60

Addition Worksheet 0 - 5

1) $2 + 3 =$ _____

2) $3 + 1 =$ _____

3) $1 + 2 =$ _____

4) $1 + 4 =$ _____

5) $0 + 4 =$ _____

6) $1 + 1 =$ _____

7) $0 + 3 =$ _____

8) $5 + 1 =$ _____

9) $3 + 2 =$ _____

10) $0 + 5 =$ _____

11) $1 + 4 =$ _____

12) $3 + 1 =$ _____

13) $1 + 2 =$ _____

14) $5 + 0 =$ _____

15) $1 + 4 =$ _____

16) $2 + 2 =$ _____

17) $2 + 3 =$ _____

18) $5 + 0 =$ _____

19) $0 + 5 =$ _____

20) $2 + 1 =$ _____

21) $4 + 1 =$ _____

22) $0 + 3 =$ _____

23) $2 + 1 =$ _____

24) $2 + 2 =$ _____

25) $2 + 2 =$ _____

26) $0 + 2 =$ _____

27) $1 + 4 =$ _____

28) $4 + 0 =$ _____

29) $1 + 4 =$ _____

30) $0 + 5 =$ _____

31) $1 + 4 =$ _____

32) $1 + 4 =$ _____

33) $1 + 3 =$ _____

34) $4 + 1 =$ _____

35) $2 + 2 =$ _____

36) $5 + 0 =$ _____

37) $4 + 1 =$ _____

38) $0 + 3 =$ _____

39) $3 + 2 =$ _____

40) $2 + 1 =$ _____

41) $1 + 3 =$ _____

42) $0 + 5 =$ _____

43) $0 + 5 =$ _____

44) $0 + 5 =$ _____

45) $4 + 1 =$ _____

46) $1 + 2 =$ _____

47) $0 + 4 =$ _____

48) $1 + 3 =$ _____

49) $1 + 4 =$ _____

50) $1 + 3 =$ _____

51) $3 + 2 =$ _____

52) $2 + 3 =$ _____

53) $1 + 1 =$ _____

54) $0 + 5 =$ _____

55) $1 + 5 =$ _____

56) $4 + 1 =$ _____

57) $4 + 0 =$ _____

58) $2 + 3 =$ _____

59) $0 + 5 =$ _____

60) $3 + 1 =$ _____

Answer Key
Addition Worksheet 0 - 5

1) 2 + 3 = **5**	2) 3 + 1 = **4**	3) 1 + 2 = **3**
4) 1 + 4 = **5**	5) 0 + 4 = **4**	6) 1 + 1 = **2**
7) 0 + 3 = **3**	8) 5 + 1 = **6**	9) 3 + 2 = **5**
10) 0 + 5 = **5**	11) 1 + 4 = **5**	12) 3 + 1 = **4**
13) 1 + 2 = **3**	14) 5 + 0 = **5**	15) 1 + 4 = **5**
16) 2 + 2 = **4**	17) 2 + 3 = **5**	18) 5 + 0 = **5**
19) 0 + 5 = **5**	20) 2 + 1 = **3**	21) 4 + 1 = **5**
22) 0 + 3 = **3**	23) 2 + 1 = **3**	24) 2 + 2 = **4**
25) 2 + 2 = **4**	26) 0 + 2 = **2**	27) 1 + 4 = **5**
28) 4 + 0 = **4**	29) 1 + 4 = **5**	30) 0 + 5 = **5**
31) 1 + 4 = **5**	32) 1 + 4 = **5**	33) 1 + 3 = **4**
34) 4 + 1 = **5**	35) 2 + 2 = **4**	36) 5 + 0 = **5**
37) 4 + 1 = **5**	38) 0 + 3 = **3**	39) 3 + 2 = **5**
40) 2 + 1 = **3**	41) 1 + 3 = **4**	42) 0 + 5 = **5**
43) 0 + 5 = **5**	44) 0 + 5 = **5**	45) 4 + 1 = **5**
46) 1 + 2 = **3**	47) 0 + 4 = **4**	48) 1 + 3 = **4**
49) 1 + 4 = **5**	50) 1 + 3 = **4**	51) 3 + 2 = **5**
52) 2 + 3 = **5**	53) 1 + 1 = **2**	54) 0 + 5 = **5**
55) 1 + 5 = **6**	56) 4 + 1 = **5**	57) 4 + 0 = **4**
58) 2 + 3 = **5**	59) 0 + 5 = **5**	60) 3 + 1 = **4**

Date : _________

Time : _________

Name : __________

Score : _____/60

Addition Worksheet 0 - 5

1) 3 + 2 = _____

2) 2 + 1 = _____

3) 3 + 2 = _____

4) 1 + 1 = _____

5) 1 + 1 = _____

6) 1 + 3 = _____

7) 1 + 2 = _____

8) 4 + 0 = _____

9) 3 + 1 = _____

10) 1 + 2 = _____

11) 5 + 0 = _____

12) 2 + 3 = _____

13) 2 + 1 = _____

14) 2 + 2 = _____

15) 1 + 4 = _____

16) 4 + 2 = _____

17) 2 + 3 = _____

18) 0 + 2 = _____

19) 3 + 2 = _____

20) 3 + 1 = _____

21) 1 + 2 = _____

22) 5 + 0 = _____

23) 0 + 4 = _____

24) 2 + 3 = _____

25) 2 + 2 = _____

26) 4 + 1 = _____

27) 0 + 1 = _____

28) 1 + 2 = _____

29) 2 + 3 = _____

30) 0 + 5 = _____

31) 3 + 2 = _____

32) 3 + 1 = _____

33) 0 + 5 = _____

34) 2 + 3 = _____

35) 3 + 1 = _____

36) 2 + 3 = _____

37) 1 + 2 = _____

38) 1 + 3 = _____

39) 1 + 3 = _____

40) 3 + 1 = _____

41) 4 + 1 = _____

42) 5 + 1 = _____

43) 1 + 3 = _____

44) 2 + 2 = _____

45) 1 + 1 = _____

46) 2 + 2 = _____

47) 0 + 4 = _____

48) 3 + 1 = _____

49) 5 + 0 = _____

50) 2 + 1 = _____

51) 3 + 2 = _____

52) 0 + 4 = _____

53) 2 + 3 = _____

54) 3 + 2 = _____

55) 3 + 2 = _____

56) 2 + 2 = _____

57) 0 + 4 = _____

58) 5 + 0 = _____

59) 4 + 0 = _____

60) 2 + 3 = _____

Answer Key
Addition Worksheet 0 - 5

1) 3 + 2 = **5**   2) 2 + 1 = **3**   3) 3 + 2 = **5**

4) 1 + 1 = **2**   5) 1 + 1 = **2**   6) 1 + 3 = **4**

7) 1 + 2 = **3**   8) 4 + 0 = **4**   9) 3 + 1 = **4**

10) 1 + 2 = **3**   11) 5 + 0 = **5**   12) 2 + 3 = **5**

13) 2 + 1 = **3**   14) 2 + 2 = **4**   15) 1 + 4 = **5**

16) 4 + 2 = **6**   17) 2 + 3 = **5**   18) 0 + 2 = **2**

19) 3 + 2 = **5**   20) 3 + 1 = **4**   21) 1 + 2 = **3**

22) 5 + 0 = **5**   23) 0 + 4 = **4**   24) 2 + 3 = **5**

25) 2 + 2 = **4**   26) 4 + 1 = **5**   27) 0 + 1 = **1**

28) 1 + 2 = **3**   29) 2 + 3 = **5**   30) 0 + 5 = **5**

31) 3 + 2 = **5**   32) 3 + 1 = **4**   33) 0 + 5 = **5**

34) 2 + 3 = **5**   35) 3 + 1 = **4**   36) 2 + 3 = **5**

37) 1 + 2 = **3**   38) 1 + 3 = **4**   39) 1 + 3 = **4**

40) 3 + 1 = **4**   41) 4 + 1 = **5**   42) 5 + 1 = **6**

43) 1 + 3 = **4**   44) 2 + 2 = **4**   45) 1 + 1 = **2**

46) 2 + 2 = **4**   47) 0 + 4 = **4**   48) 3 + 1 = **4**

49) 5 + 0 = **5**   50) 2 + 1 = **3**   51) 3 + 2 = **5**

52) 0 + 4 = **4**   53) 2 + 3 = **5**   54) 3 + 2 = **5**

55) 3 + 2 = **5**   56) 2 + 2 = **4**   57) 0 + 4 = **4**

58) 5 + 0 = **5**   59) 4 + 0 = **4**   60) 2 + 3 = **5**

Addition Worksheet 0 - 5

1) $4 + 1 =$ _____

2) $2 + 2 =$ _____

3) $2 + 3 =$ _____

4) $2 + 2 =$ _____

5) $5 + 1 =$ _____

6) $4 + 1 =$ _____

7) $0 + 0 =$ _____

8) $5 + 0 =$ _____

9) $3 + 1 =$ _____

10) $2 + 2 =$ _____

11) $1 + 1 =$ _____

12) $2 + 3 =$ _____

13) $1 + 2 =$ _____

14) $2 + 2 =$ _____

15) $1 + 4 =$ _____

16) $1 + 4 =$ _____

17) $2 + 2 =$ _____

18) $1 + 3 =$ _____

19) $5 + 0 =$ _____

20) $1 + 1 =$ _____

21) $2 + 2 =$ _____

22) $2 + 2 =$ _____

23) $1 + 1 =$ _____

24) $5 + 0 =$ _____

25) $2 + 1 =$ _____

26) $5 + 0 =$ _____

27) $1 + 3 =$ _____

28) $1 + 3 =$ _____

29) $0 + 5 =$ _____

30) $4 + 0 =$ _____

31) $4 + 1 =$ _____

32) $0 + 5 =$ _____

33) $3 + 1 =$ _____

34) $2 + 1 =$ _____

35) $1 + 3 =$ _____

36) $0 + 1 =$ _____

37) $2 + 3 =$ _____

38) $3 + 2 =$ _____

39) $2 + 2 =$ _____

40) $0 + 5 =$ _____

41) $5 + 0 =$ _____

42) $2 + 0 =$ _____

43) $2 + 3 =$ _____

44) $1 + 2 =$ _____

45) $1 + 4 =$ _____

46) $0 + 5 =$ _____

47) $0 + 2 =$ _____

48) $2 + 2 =$ _____

49) $1 + 3 =$ _____

50) $2 + 3 =$ _____

51) $1 + 1 =$ _____

52) $3 + 2 =$ _____

53) $1 + 1 =$ _____

54) $2 + 3 =$ _____

55) $2 + 2 =$ _____

56) $1 + 3 =$ _____

57) $1 + 1 =$ _____

58) $4 + 1 =$ _____

59) $0 + 1 =$ _____

60) $1 + 4 =$ _____

Answer Key
Addition Worksheet 0 - 5

1) 4 + 1 = **5**

2) 2 + 2 = **4**

3) 2 + 3 = **5**

4) 2 + 2 = **4**

5) 5 + 1 = **6**

6) 4 + 1 = **5**

7) 0 + 0 = **0**

8) 5 + 0 = **5**

9) 3 + 1 = **4**

10) 2 + 2 = **4**

11) 1 + 1 = **2**

12) 2 + 3 = **5**

13) 1 + 2 = **3**

14) 2 + 2 = **4**

15) 1 + 4 = **5**

16) 1 + 4 = **5**

17) 2 + 2 = **4**

18) 1 + 3 = **4**

19) 5 + 0 = **5**

20) 1 + 1 = **2**

21) 2 + 2 = **4**

22) 2 + 2 = **4**

23) 1 + 1 = **2**

24) 5 + 0 = **5**

25) 2 + 1 = **3**

26) 5 + 0 = **5**

27) 1 + 3 = **4**

28) 1 + 3 = **4**

29) 0 + 5 = **5**

30) 4 + 0 = **4**

31) 4 + 1 = **5**

32) 0 + 5 = **5**

33) 3 + 1 = **4**

34) 2 + 1 = **3**

35) 1 + 3 = **4**

36) 0 + 1 = **1**

37) 2 + 3 = **5**

38) 3 + 2 = **5**

39) 2 + 2 = **4**

40) 0 + 5 = **5**

41) 5 + 0 = **5**

42) 2 + 0 = **2**

43) 2 + 3 = **5**

44) 1 + 2 = **3**

45) 1 + 4 = **5**

46) 0 + 5 = **5**

47) 0 + 2 = **2**

48) 2 + 2 = **4**

49) 1 + 3 = **4**

50) 2 + 3 = **5**

51) 1 + 1 = **2**

52) 3 + 2 = **5**

53) 1 + 1 = **2**

54) 2 + 3 = **5**

55) 2 + 2 = **4**

56) 1 + 3 = **4**

57) 1 + 1 = **2**

58) 4 + 1 = **5**

59) 0 + 1 = **1**

60) 1 + 4 = **5**

Addition Worksheet 0 - 5

1) 1 + 2 = _____	2) 5 + 0 = _____	3) 2 + 3 = _____
4) 3 + 3 = _____	5) 0 + 4 = _____	6) 3 + 2 = _____
7) 2 + 2 = _____	8) 2 + 1 = _____	9) 1 + 2 = _____
10) 1 + 4 = _____	11) 2 + 2 = _____	12) 1 + 3 = _____
13) 1 + 1 = _____	14) 1 + 3 = _____	15) 1 + 2 = _____
16) 2 + 2 = _____	17) 0 + 5 = _____	18) 2 + 2 = _____
19) 0 + 3 = _____	20) 3 + 0 = _____	21) 0 + 4 = _____
22) 1 + 3 = _____	23) 3 + 2 = _____	24) 5 + 0 = _____
25) 1 + 4 = _____	26) 0 + 5 = _____	27) 2 + 3 = _____
28) 3 + 2 = _____	29) 0 + 5 = _____	30) 3 + 2 = _____
31) 1 + 3 = _____	32) 3 + 1 = _____	33) 1 + 3 = _____
34) 1 + 3 = _____	35) 3 + 2 = _____	36) 5 + 0 = _____
37) 0 + 5 = _____	38) 3 + 2 = _____	39) 0 + 3 = _____
40) 0 + 2 = _____	41) 3 + 0 = _____	42) 0 + 5 = _____
43) 2 + 1 = _____	44) 0 + 4 = _____	45) 0 + 5 = _____
46) 1 + 2 = _____	47) 2 + 3 = _____	48) 2 + 2 = _____
49) 4 + 1 = _____	50) 2 + 1 = _____	51) 1 + 2 = _____
52) 2 + 1 = _____	53) 1 + 4 = _____	54) 2 + 2 = _____
55) 2 + 0 = _____	56) 1 + 4 = _____	57) 3 + 1 = _____
58) 3 + 2 = _____	59) 1 + 3 = _____	60) 5 + 0 = _____

Answer Key
Addition Worksheet 0 - 5

1) 1 + 2 = **3**

2) 5 + 0 = **5**

3) 2 + 3 = **5**

4) 3 + 3 = **6**

5) 0 + 4 = **4**

6) 3 + 2 = **5**

7) 2 + 2 = **4**

8) 2 + 1 = **3**

9) 1 + 2 = **3**

10) 1 + 4 = **5**

11) 2 + 2 = **4**

12) 1 + 3 = **4**

13) 1 + 1 = **2**

14) 1 + 3 = **4**

15) 1 + 2 = **3**

16) 2 + 2 = **4**

17) 0 + 5 = **5**

18) 2 + 2 = **4**

19) 0 + 3 = **3**

20) 3 + 0 = **3**

21) 0 + 4 = **4**

22) 1 + 3 = **4**

23) 3 + 2 = **5**

24) 5 + 0 = **5**

25) 1 + 4 = **5**

26) 0 + 5 = **5**

27) 2 + 3 = **5**

28) 3 + 2 = **5**

29) 0 + 5 = **5**

30) 3 + 2 = **5**

31) 1 + 3 = **4**

32) 3 + 1 = **4**

33) 1 + 3 = **4**

34) 1 + 3 = **4**

35) 3 + 2 = **5**

36) 5 + 0 = **5**

37) 0 + 5 = **5**

38) 3 + 2 = **5**

39) 0 + 3 = **3**

40) 0 + 2 = **2**

41) 3 + 0 = **3**

42) 0 + 5 = **5**

43) 2 + 1 = **3**

44) 0 + 4 = **4**

45) 0 + 5 = **5**

46) 1 + 2 = **3**

47) 2 + 3 = **5**

48) 2 + 2 = **4**

49) 4 + 1 = **5**

50) 2 + 1 = **3**

51) 1 + 2 = **3**

52) 2 + 1 = **3**

53) 1 + 4 = **5**

54) 2 + 2 = **4**

55) 2 + 0 = **2**

56) 1 + 4 = **5**

57) 3 + 1 = **4**

58) 3 + 2 = **5**

59) 1 + 3 = **4**

60) 5 + 0 = **5**

Date : _________   Name : __________

Time : _________   Score : _____/60

Addition Worksheet 0 - 10

1) 0 + 10 = _____   2) 3 + 5 = _____   3) 9 + 0 = _____

4) 10 + 0 = _____   5) 1 + 6 = _____   6) 3 + 1 = _____

7) 8 + 1 = _____   8) 2 + 8 = _____   9) 10 + 0 = _____

10) 4 + 1 = _____   11) 1 + 7 = _____   12) 2 + 8 = _____

13) 9 + 1 = _____   14) 1 + 4 = _____   15) 6 + 5 = _____

16) 2 + 3 = _____   17) 1 + 7 = _____   18) 0 + 10 = _____

19) 6 + 3 = _____   20) 8 + 1 = _____   21) 3 + 6 = _____

22) 6 + 3 = _____   23) 3 + 7 = _____   24) 3 + 4 = _____

25) 9 + 1 = _____   26) 0 + 5 = _____   27) 6 + 1 = _____

28) 1 + 9 = _____   29) 4 + 4 = _____   30) 5 + 4 = _____

31) 3 + 7 = _____   32) 1 + 3 = _____   33) 3 + 4 = _____

34) 8 + 2 = _____   35) 8 + 1 = _____   36) 3 + 2 = _____

37) 5 + 4 = _____   38) 8 + 1 = _____   39) 10 + 0 = _____

40) 5 + 0 = _____   41) 8 + 3 = _____   42) 0 + 10 = _____

43) 2 + 1 = _____   44) 4 + 4 = _____   45) 1 + 7 = _____

46) 5 + 2 = _____   47) 2 + 4 = _____   48) 8 + 2 = _____

49) 1 + 5 = _____   50) 1 + 8 = _____   51) 1 + 10 = _____

52) 6 + 1 = _____   53) 2 + 3 = _____   54) 3 + 4 = _____

55) 2 + 7 = _____   56) 3 + 5 = _____   57) 8 + 2 = _____

58) 2 + 6 = _____   59) 3 + 6 = _____   60) 4 + 5 = _____

Answer Key
Addition Worksheet 0 - 10

1) 0 + 10 = **10**	2) 3 + 5 = **8**	3) 9 + 0 = **9**
4) 10 + 0 = **10**	5) 1 + 6 = **7**	6) 3 + 1 = **4**
7) 8 + 1 = **9**	8) 2 + 8 = **10**	9) 10 + 0 = **10**
10) 4 + 1 = **5**	11) 1 + 7 = **8**	12) 2 + 8 = **10**
13) 9 + 1 = **10**	14) 1 + 4 = **5**	15) 6 + 5 = **11**
16) 2 + 3 = **5**	17) 1 + 7 = **8**	18) 0 + 10 = **10**
19) 6 + 3 = **9**	20) 8 + 1 = **9**	21) 3 + 6 = **9**
22) 6 + 3 = **9**	23) 3 + 7 = **10**	24) 3 + 4 = **7**
25) 9 + 1 = **10**	26) 0 + 5 = **5**	27) 6 + 1 = **7**
28) 1 + 9 = **10**	29) 4 + 4 = **8**	30) 5 + 4 = **9**
31) 3 + 7 = **10**	32) 1 + 3 = **4**	33) 3 + 4 = **7**
34) 8 + 2 = **10**	35) 8 + 1 = **9**	36) 3 + 2 = **5**
37) 5 + 4 = **9**	38) 8 + 1 = **9**	39) 10 + 0 = **10**
40) 5 + 0 = **5**	41) 8 + 3 = **11**	42) 0 + 10 = **10**
43) 2 + 1 = **3**	44) 4 + 4 = **8**	45) 1 + 7 = **8**
46) 5 + 2 = **7**	47) 2 + 4 = **6**	48) 8 + 2 = **10**
49) 1 + 5 = **6**	50) 1 + 8 = **9**	51) 1 + 10 = **11**
52) 6 + 1 = **7**	53) 2 + 3 = **5**	54) 3 + 4 = **7**
55) 2 + 7 = **9**	56) 3 + 5 = **8**	57) 8 + 2 = **10**
58) 2 + 6 = **8**	59) 3 + 6 = **9**	60) 4 + 5 = **9**

Date : _________

Time : _________

Name : __________

Score : _____/60

Addition Worksheet 0 - 10

1) 1 + 9 = _____

2) 3 + 0 = _____

3) 7 + 3 = _____

4) 3 + 4 = _____

5) 5 + 3 = _____

6) 5 + 3 = _____

7) 8 + 1 = _____

8) 2 + 4 = _____

9) 0 + 5 = _____

10) 6 + 2 = _____

11) 7 + 3 = _____

12) 5 + 4 = _____

13) 4 + 1 = _____

14) 2 + 6 = _____

15) 5 + 2 = _____

16) 4 + 3 = _____

17) 1 + 4 = _____

18) 8 + 1 = _____

19) 7 + 2 = _____

20) 7 + 0 = _____

21) 1 + 9 = _____

22) 5 + 2 = _____

23) 1 + 9 = _____

24) 8 + 2 = _____

25) 8 + 1 = _____

26) 1 + 4 = _____

27) 0 + 10 = _____

28) 5 + 0 = _____

29) 4 + 4 = _____

30) 2 + 8 = _____

31) 4 + 2 = _____

32) 3 + 1 = _____

33) 5 + 2 = _____

34) 6 + 3 = _____

35) 5 + 4 = _____

36) 0 + 10 = _____

37) 1 + 7 = _____

38) 3 + 4 = _____

39) 10 + 0 = _____

40) 3 + 3 = _____

41) 5 + 1 = _____

42) 2 + 5 = _____

43) 10 + 1 = _____

44) 6 + 4 = _____

45) 7 + 3 = _____

46) 1 + 5 = _____

47) 2 + 6 = _____

48) 6 + 1 = _____

49) 5 + 1 = _____

50) 1 + 7 = _____

51) 0 + 4 = _____

52) 2 + 4 = _____

53) 5 + 1 = _____

54) 2 + 5 = _____

55) 4 + 6 = _____

56) 1 + 7 = _____

57) 3 + 3 = _____

58) 2 + 4 = _____

59) 1 + 4 = _____

60) 8 + 2 = _____

Answer Key
Addition Worksheet 0 - 10

1) 1 + 9 = **10**

2) 3 + 0 = **3**

3) 7 + 3 = **10**

4) 3 + 4 = **7**

5) 5 + 3 = **8**

6) 5 + 3 = **8**

7) 8 + 1 = **9**

8) 2 + 4 = **6**

9) 0 + 5 = **5**

10) 6 + 2 = **8**

11) 7 + 3 = **10**

12) 5 + 4 = **9**

13) 4 + 1 = **5**

14) 2 + 6 = **8**

15) 5 + 2 = **7**

16) 4 + 3 = **7**

17) 1 + 4 = **5**

18) 8 + 1 = **9**

19) 7 + 2 = **9**

20) 7 + 0 = **7**

21) 1 + 9 = **10**

22) 5 + 2 = **7**

23) 1 + 9 = **10**

24) 8 + 2 = **10**

25) 8 + 1 = **9**

26) 1 + 4 = **5**

27) 0 + 10 = **10**

28) 5 + 0 = **5**

29) 4 + 4 = **8**

30) 2 + 8 = **10**

31) 4 + 2 = **6**

32) 3 + 1 = **4**

33) 5 + 2 = **7**

34) 6 + 3 = **9**

35) 5 + 4 = **9**

36) 0 + 10 = **10**

37) 1 + 7 = **8**

38) 3 + 4 = **7**

39) 10 + 0 = **10**

40) 3 + 3 = **6**

41) 5 + 1 = **6**

42) 2 + 5 = **7**

43) 10 + 1 = **11**

44) 6 + 4 = **10**

45) 7 + 3 = **10**

46) 1 + 5 = **6**

47) 2 + 6 = **8**

48) 6 + 1 = **7**

49) 5 + 1 = **6**

50) 1 + 7 = **8**

51) 0 + 4 = **4**

52) 2 + 4 = **6**

53) 5 + 1 = **6**

54) 2 + 5 = **7**

55) 4 + 6 = **10**

56) 1 + 7 = **8**

57) 3 + 3 = **6**

58) 2 + 4 = **6**

59) 1 + 4 = **5**

60) 8 + 2 = **10**

Date : _________   Name : __________

Time : _______   Score : ______/60

Addition Worksheet 0 - 10

1) 1 + 7 = _____   2) 8 + 2 = _____   3) 8 + 2 = _____

4) 9 + 1 = _____   5) 5 + 3 = _____   6) 4 + 6 = _____

7) 3 + 3 = _____   8) 1 + 7 = _____   9) 5 + 2 = _____

10) 9 + 1 = _____   11) 7 + 3 = _____   12) 1 + 9 = _____

13) 7 + 1 = _____   14) 2 + 9 = _____   15) 6 + 2 = _____

16) 3 + 3 = _____   17) 3 + 6 = _____   18) 5 + 3 = _____

19) 4 + 4 = _____   20) 3 + 1 = _____   21) 3 + 7 = _____

22) 10 + 0 = _____   23) 1 + 7 = _____   24) 1 + 5 = _____

25) 9 + 1 = _____   26) 0 + 1 = _____   27) 3 + 1 = _____

28) 9 + 1 = _____   29) 5 + 5 = _____   30) 7 + 1 = _____

31) 1 + 6 = _____   32) 3 + 2 = _____   33) 3 + 6 = _____

34) 4 + 6 = _____   35) 6 + 3 = _____   36) 3 + 5 = _____

37) 1 + 7 = _____   38) 4 + 4 = _____   39) 4 + 0 = _____

40) 6 + 2 = _____   41) 1 + 2 = _____   42) 3 + 5 = _____

43) 5 + 4 = _____   44) 0 + 10 = _____   45) 3 + 3 = _____

46) 6 + 3 = _____   47) 0 + 10 = _____   48) 10 + 0 = _____

49) 7 + 2 = _____   50) 6 + 3 = _____   51) 0 + 10 = _____

52) 8 + 1 = _____   53) 5 + 3 = _____   54) 5 + 3 = _____

55) 3 + 4 = _____   56) 2 + 8 = _____   57) 1 + 3 = _____

58) 1 + 7 = _____   59) 10 + 0 = _____   60) 4 + 6 = _____

Answer Key
Addition Worksheet 0 - 10

1) $1 + 7 = \mathbf{8}$	2) $8 + 2 = \mathbf{10}$	3) $8 + 2 = \mathbf{10}$
4) $9 + 1 = \mathbf{10}$	5) $5 + 3 = \mathbf{8}$	6) $4 + 6 = \mathbf{10}$
7) $3 + 3 = \mathbf{6}$	8) $1 + 7 = \mathbf{8}$	9) $5 + 2 = \mathbf{7}$
10) $9 + 1 = \mathbf{10}$	11) $7 + 3 = \mathbf{10}$	12) $1 + 9 = \mathbf{10}$
13) $7 + 1 = \mathbf{8}$	14) $2 + 9 = \mathbf{11}$	15) $6 + 2 = \mathbf{8}$
16) $3 + 3 = \mathbf{6}$	17) $3 + 6 = \mathbf{9}$	18) $5 + 3 = \mathbf{8}$
19) $4 + 4 = \mathbf{8}$	20) $3 + 1 = \mathbf{4}$	21) $3 + 7 = \mathbf{10}$
22) $10 + 0 = \mathbf{10}$	23) $1 + 7 = \mathbf{8}$	24) $1 + 5 = \mathbf{6}$
25) $9 + 1 = \mathbf{10}$	26) $0 + 1 = \mathbf{1}$	27) $3 + 1 = \mathbf{4}$
28) $9 + 1 = \mathbf{10}$	29) $5 + 5 = \mathbf{10}$	30) $7 + 1 = \mathbf{8}$
31) $1 + 6 = \mathbf{7}$	32) $3 + 2 = \mathbf{5}$	33) $3 + 6 = \mathbf{9}$
34) $4 + 6 = \mathbf{10}$	35) $6 + 3 = \mathbf{9}$	36) $3 + 5 = \mathbf{8}$
37) $1 + 7 = \mathbf{8}$	38) $4 + 4 = \mathbf{8}$	39) $4 + 0 = \mathbf{4}$
40) $6 + 2 = \mathbf{8}$	41) $1 + 2 = \mathbf{3}$	42) $3 + 5 = \mathbf{8}$
43) $5 + 4 = \mathbf{9}$	44) $0 + 10 = \mathbf{10}$	45) $3 + 3 = \mathbf{6}$
46) $6 + 3 = \mathbf{9}$	47) $0 + 10 = \mathbf{10}$	48) $10 + 0 = \mathbf{10}$
49) $7 + 2 = \mathbf{9}$	50) $6 + 3 = \mathbf{9}$	51) $0 + 10 = \mathbf{10}$
52) $8 + 1 = \mathbf{9}$	53) $5 + 3 = \mathbf{8}$	54) $5 + 3 = \mathbf{8}$
55) $3 + 4 = \mathbf{7}$	56) $2 + 8 = \mathbf{10}$	57) $1 + 3 = \mathbf{4}$
58) $1 + 7 = \mathbf{8}$	59) $10 + 0 = \mathbf{10}$	60) $4 + 6 = \mathbf{10}$

Addition Worksheet 0 - 10

1) 6 + 1 = _____	2) 3 + 6 = _____	3) 1 + 9 = _____
4) 7 + 3 = _____	5) 3 + 1 = _____	6) 1 + 6 = _____
7) 4 + 3 = _____	8) 8 + 2 = _____	9) 1 + 3 = _____
10) 0 + 10 = _____	11) 4 + 1 = _____	12) 3 + 6 = _____
13) 9 + 1 = _____	14) 5 + 4 = _____	15) 6 + 1 = _____
16) 10 + 0 = _____	17) 4 + 2 = _____	18) 0 + 8 = _____
19) 2 + 8 = _____	20) 8 + 1 = _____	21) 9 + 0 = _____
22) 1 + 3 = _____	23) 2 + 3 = _____	24) 7 + 1 = _____
25) 3 + 2 = _____	26) 6 + 1 = _____	27) 3 + 3 = _____
28) 2 + 1 = _____	29) 2 + 7 = _____	30) 10 + 0 = _____
31) 5 + 2 = _____	32) 5 + 1 = _____	33) 3 + 1 = _____
34) 1 + 3 = _____	35) 4 + 1 = _____	36) 4 + 3 = _____
37) 1 + 1 = _____	38) 3 + 6 = _____	39) 1 + 5 = _____
40) 6 + 3 = _____	41) 6 + 3 = _____	42) 9 + 1 = _____
43) 8 + 1 = _____	44) 7 + 3 = _____	45) 1 + 9 = _____
46) 7 + 3 = _____	47) 7 + 3 = _____	48) 10 + 1 = _____
49) 0 + 9 = _____	50) 8 + 1 = _____	51) 6 + 4 = _____
52) 3 + 3 = _____	53) 5 + 2 = _____	54) 4 + 4 = _____
55) 1 + 4 = _____	56) 3 + 7 = _____	57) 1 + 7 = _____
58) 7 + 2 = _____	59) 4 + 6 = _____	60) 4 + 4 = _____

Answer Key
Addition Worksheet 0 - 10

1) 6 + 1 = **7**	2) 3 + 6 = **9**	3) 1 + 9 = **10**
4) 7 + 3 = **10**	5) 3 + 1 = **4**	6) 1 + 6 = **7**
7) 4 + 3 = **7**	8) 8 + 2 = **10**	9) 1 + 3 = **4**
10) 0 + 10 = **10**	11) 4 + 1 = **5**	12) 3 + 6 = **9**
13) 9 + 1 = **10**	14) 5 + 4 = **9**	15) 6 + 1 = **7**
16) 10 + 0 = **10**	17) 4 + 2 = **6**	18) 0 + 8 = **8**
19) 2 + 8 = **10**	20) 8 + 1 = **9**	21) 9 + 0 = **9**
22) 1 + 3 = **4**	23) 2 + 3 = **5**	24) 7 + 1 = **8**
25) 3 + 2 = **5**	26) 6 + 1 = **7**	27) 3 + 3 = **6**
28) 2 + 1 = **3**	29) 2 + 7 = **9**	30) 10 + 0 = **10**
31) 5 + 2 = **7**	32) 5 + 1 = **6**	33) 3 + 1 = **4**
34) 1 + 3 = **4**	35) 4 + 1 = **5**	36) 4 + 3 = **7**
37) 1 + 1 = **2**	38) 3 + 6 = **9**	39) 1 + 5 = **6**
40) 6 + 3 = **9**	41) 6 + 3 = **9**	42) 9 + 1 = **10**
43) 8 + 1 = **9**	44) 7 + 3 = **10**	45) 1 + 9 = **10**
46) 7 + 3 = **10**	47) 7 + 3 = **10**	48) 10 + 1 = **11**
49) 0 + 9 = **9**	50) 8 + 1 = **9**	51) 6 + 4 = **10**
52) 3 + 3 = **6**	53) 5 + 2 = **7**	54) 4 + 4 = **8**
55) 1 + 4 = **5**	56) 3 + 7 = **10**	57) 1 + 7 = **8**
58) 7 + 2 = **9**	59) 4 + 6 = **10**	60) 4 + 4 = **8**

Date : _________   Name : _________

Time : _________   Score : _____/60

Addition Worksheet 0 - 10

1) 8 + 2 = _____	2) 7 + 1 = _____	3) 1 + 4 = _____
4) 1 + 9 = _____	5) 5 + 3 = _____	6) 1 + 7 = _____
7) 1 + 8 = _____	8) 9 + 1 = _____	9) 4 + 5 = _____
10) 9 + 1 = _____	11) 3 + 6 = _____	12) 4 + 3 = _____
13) 4 + 3 = _____	14) 10 + 0 = _____	15) 7 + 2 = _____
16) 9 + 1 = _____	17) 0 + 10 = _____	18) 7 + 2 = _____
19) 5 + 2 = _____	20) 3 + 5 = _____	21) 0 + 10 = _____
22) 6 + 2 = _____	23) 2 + 8 = _____	24) 1 + 8 = _____
25) 1 + 4 = _____	26) 7 + 1 = _____	27) 6 + 1 = _____
28) 3 + 4 = _____	29) 4 + 4 = _____	30) 8 + 1 = _____
31) 7 + 3 = _____	32) 1 + 8 = _____	33) 6 + 1 = _____
34) 3 + 6 = _____	35) 3 + 1 = _____	36) 4 + 3 = _____
37) 3 + 6 = _____	38) 0 + 9 = _____	39) 5 + 2 = _____
40) 2 + 3 = _____	41) 3 + 6 = _____	42) 1 + 8 = _____
43) 7 + 2 = _____	44) 3 + 7 = _____	45) 3 + 3 = _____
46) 5 + 1 = _____	47) 10 + 0 = _____	48) 6 + 2 = _____
49) 8 + 2 = _____	50) 4 + 6 = _____	51) 0 + 7 = _____
52) 6 + 1 = _____	53) 1 + 9 = _____	54) 1 + 6 = _____
55) 2 + 5 = _____	56) 7 + 1 = _____	57) 4 + 2 = _____
58) 3 + 5 = _____	59) 10 + 0 = _____	60) 6 + 1 = _____

Answer Key

Addition Worksheet 0 - 10

1) 8 + 2 = **10**

2) 7 + 1 = **8**

3) 1 + 4 = **5**

4) 1 + 9 = **10**

5) 5 + 3 = **8**

6) 1 + 7 = **8**

7) 1 + 8 = **9**

8) 9 + 1 = **10**

9) 4 + 5 = **9**

10) 9 + 1 = **10**

11) 3 + 6 = **9**

12) 4 + 3 = **7**

13) 4 + 3 = **7**

14) 10 + 0 = **10**

15) 7 + 2 = **9**

16) 9 + 1 = **10**

17) 0 + 10 = **10**

18) 7 + 2 = **9**

19) 5 + 2 = **7**

20) 3 + 5 = **8**

21) 0 + 10 = **10**

22) 6 + 2 = **8**

23) 2 + 8 = **10**

24) 1 + 8 = **9**

25) 1 + 4 = **5**

26) 7 + 1 = **8**

27) 6 + 1 = **7**

28) 3 + 4 = **7**

29) 4 + 4 = **8**

30) 8 + 1 = **9**

31) 7 + 3 = **10**

32) 1 + 8 = **9**

33) 6 + 1 = **7**

34) 3 + 6 = **9**

35) 3 + 1 = **4**

36) 4 + 3 = **7**

37) 3 + 6 = **9**

38) 0 + 9 = **9**

39) 5 + 2 = **7**

40) 2 + 3 = **5**

41) 3 + 6 = **9**

42) 1 + 8 = **9**

43) 7 + 2 = **9**

44) 3 + 7 = **10**

45) 3 + 3 = **6**

46) 5 + 1 = **6**

47) 10 + 0 = **10**

48) 6 + 2 = **8**

49) 8 + 2 = **10**

50) 4 + 6 = **10**

51) 0 + 7 = **7**

52) 6 + 1 = **7**

53) 1 + 9 = **10**

54) 1 + 6 = **7**

55) 2 + 5 = **7**

56) 7 + 1 = **8**

57) 4 + 2 = **6**

58) 3 + 5 = **8**

59) 10 + 0 = **10**

60) 6 + 1 = **7**

Date : _________

Time : _________

Name : __________

Score : _____/60

Addition Worksheet 0 - 20

1) 5 + 8 = _____

2) 1 + 18 = _____

3) 8 + 7 = _____

4) 2 + 18 = _____

5) 13 + 3 = _____

6) 10 + 1 = _____

7) 1 + 11 = _____

8) 18 + 2 = _____

9) 11 + 3 = _____

10) 2 + 15 = _____

11) 5 + 15 = _____

12) 11 + 4 = _____

13) 12 + 3 = _____

14) 14 + 1 = _____

15) 14 + 4 = _____

16) 1 + 15 = _____

17) 1 + 17 = _____

18) 2 + 9 = _____

19) 3 + 12 = _____

20) 11 + 7 = _____

21) 10 + 1 = _____

22) 19 + 1 = _____

23) 2 + 18 = _____

24) 14 + 1 = _____

25) 7 + 4 = _____

26) 8 + 2 = _____

27) 0 + 12 = _____

28) 0 + 7 = _____

29) 9 + 6 = _____

30) 11 + 7 = _____

31) 18 + 0 = _____

32) 13 + 6 = _____

33) 7 + 6 = _____

34) 3 + 15 = _____

35) 5 + 15 = _____

36) 13 + 1 = _____

37) 5 + 13 = _____

38) 5 + 5 = _____

39) 15 + 1 = _____

40) 12 + 7 = _____

41) 13 + 1 = _____

42) 7 + 2 = _____

43) 2 + 19 = _____

44) 9 + 3 = _____

45) 5 + 4 = _____

46) 2 + 17 = _____

47) 2 + 17 = _____

48) 1 + 17 = _____

49) 5 + 13 = _____

50) 7 + 5 = _____

51) 13 + 4 = _____

52) 3 + 3 = _____

53) 1 + 3 = _____

54) 3 + 2 = _____

55) 11 + 9 = _____

56) 14 + 2 = _____

57) 3 + 10 = _____

58) 2 + 17 = _____

59) 8 + 2 = _____

60) 2 + 7 = _____

Answer Key
Addition Worksheet 0 - 20

1) 5 + 8 = **13**	2) 1 + 18 = **19**	3) 8 + 7 = **15**
4) 2 + 18 = **20**	5) 13 + 3 = **16**	6) 10 + 1 = **11**
7) 1 + 11 = **12**	8) 18 + 2 = **20**	9) 11 + 3 = **14**
10) 2 + 15 = **17**	11) 5 + 15 = **20**	12) 11 + 4 = **15**
13) 12 + 3 = **15**	14) 14 + 1 = **15**	15) 14 + 4 = **18**
16) 1 + 15 = **16**	17) 1 + 17 = **18**	18) 2 + 9 = **11**
19) 3 + 12 = **15**	20) 11 + 7 = **18**	21) 10 + 1 = **11**
22) 19 + 1 = **20**	23) 2 + 18 = **20**	24) 14 + 1 = **15**
25) 7 + 4 = **11**	26) 8 + 2 = **10**	27) 0 + 12 = **12**
28) 0 + 7 = **7**	29) 9 + 6 = **15**	30) 11 + 7 = **18**
31) 18 + 0 = **18**	32) 13 + 6 = **19**	33) 7 + 6 = **13**
34) 3 + 15 = **18**	35) 5 + 15 = **20**	36) 13 + 1 = **14**
37) 5 + 13 = **18**	38) 5 + 5 = **10**	39) 15 + 1 = **16**
40) 12 + 7 = **19**	41) 13 + 1 = **14**	42) 7 + 2 = **9**
43) 2 + 19 = **21**	44) 9 + 3 = **12**	45) 5 + 4 = **9**
46) 2 + 17 = **19**	47) 2 + 17 = **19**	48) 1 + 17 = **18**
49) 5 + 13 = **18**	50) 7 + 5 = **12**	51) 13 + 4 = **17**
52) 3 + 3 = **6**	53) 1 + 3 = **4**	54) 3 + 2 = **5**
55) 11 + 9 = **20**	56) 14 + 2 = **16**	57) 3 + 10 = **13**
58) 2 + 17 = **19**	59) 8 + 2 = **10**	60) 2 + 7 = **9**

Addition Worksheet 0 - 20

1) 8 + 1 = _____

2) 2 + 5 = _____

3) 3 + 3 = _____

4) 3 + 8 = _____

5) 1 + 19 = _____

6) 2 + 4 = _____

7) 1 + 19 = _____

8) 11 + 2 = _____

9) 2 + 7 = _____

10) 3 + 13 = _____

11) 14 + 2 = _____

12) 18 + 1 = _____

13) 5 + 14 = _____

14) 0 + 19 = _____

15) 2 + 12 = _____

16) 13 + 2 = _____

17) 5 + 14 = _____

18) 2 + 3 = _____

19) 5 + 1 = _____

20) 13 + 7 = _____

21) 11 + 5 = _____

22) 4 + 5 = _____

23) 5 + 8 = _____

24) 1 + 16 = _____

25) 13 + 6 = _____

26) 1 + 17 = _____

27) 0 + 20 = _____

28) 14 + 1 = _____

29) 4 + 11 = _____

30) 5 + 5 = _____

31) 8 + 7 = _____

32) 10 + 6 = _____

33) 7 + 2 = _____

34) 1 + 7 = _____

35) 13 + 6 = _____

36) 2 + 18 = _____

37) 9 + 3 = _____

38) 9 + 10 = _____

39) 2 + 4 = _____

40) 15 + 0 = _____

41) 1 + 11 = _____

42) 12 + 8 = _____

43) 13 + 3 = _____

44) 19 + 1 = _____

45) 10 + 7 = _____

46) 5 + 0 = _____

47) 4 + 9 = _____

48) 15 + 5 = _____

49) 2 + 14 = _____

50) 10 + 11 = _____

51) 12 + 1 = _____

52) 3 + 2 = _____

53) 20 + 0 = _____

54) 2 + 7 = _____

55) 2 + 13 = _____

56) 1 + 19 = _____

57) 13 + 4 = _____

58) 1 + 10 = _____

59) 10 + 6 = _____

60) 3 + 17 = _____

# Answer Key

Addition Worksheet 0 - 20

1) 8 + 1 = **9**	2) 2 + 5 = **7**	3) 3 + 3 = **6**
4) 3 + 8 = **11**	5) 1 + 19 = **20**	6) 2 + 4 = **6**
7) 1 + 19 = **20**	8) 11 + 2 = **13**	9) 2 + 7 = **9**
10) 3 + 13 = **16**	11) 14 + 2 = **16**	12) 18 + 1 = **19**
13) 5 + 14 = **19**	14) 0 + 19 = **19**	15) 2 + 12 = **14**
16) 13 + 2 = **15**	17) 5 + 14 = **19**	18) 2 + 3 = **5**
19) 5 + 1 = **6**	20) 13 + 7 = **20**	21) 11 + 5 = **16**
22) 4 + 5 = **9**	23) 5 + 8 = **13**	24) 1 + 16 = **17**
25) 13 + 6 = **19**	26) 1 + 17 = **18**	27) 0 + 20 = **20**
28) 14 + 1 = **15**	29) 4 + 11 = **15**	30) 5 + 5 = **10**
31) 8 + 7 = **15**	32) 10 + 6 = **16**	33) 7 + 2 = **9**
34) 1 + 7 = **8**	35) 13 + 6 = **19**	36) 2 + 18 = **20**
37) 9 + 3 = **12**	38) 9 + 10 = **19**	39) 2 + 4 = **6**
40) 15 + 0 = **15**	41) 1 + 11 = **12**	42) 12 + 8 = **20**
43) 13 + 3 = **16**	44) 19 + 1 = **20**	45) 10 + 7 = **17**
46) 5 + 0 = **5**	47) 4 + 9 = **13**	48) 15 + 5 = **20**
49) 2 + 14 = **16**	50) 10 + 11 = **21**	51) 12 + 1 = **13**
52) 3 + 2 = **5**	53) 20 + 0 = **20**	54) 2 + 7 = **9**
55) 2 + 13 = **15**	56) 1 + 19 = **20**	57) 13 + 4 = **17**
58) 1 + 10 = **11**	59) 10 + 6 = **16**	60) 3 + 17 = **20**

Date : _________

Time : _________

Name : _________

Score : _____/60

Addition Worksheet 0 - 20

1) 1 + 17 = _____

2) 5 + 13 = _____

3) 3 + 8 = _____

4) 17 + 1 = _____

5) 20 + 0 = _____

6) 5 + 10 = _____

7) 8 + 6 = _____

8) 12 + 6 = _____

9) 3 + 11 = _____

10) 4 + 16 = _____

11) 10 + 9 = _____

12) 12 + 2 = _____

13) 3 + 14 = _____

14) 4 + 12 = _____

15) 13 + 5 = _____

16) 3 + 14 = _____

17) 14 + 2 = _____

18) 5 + 2 = _____

19) 18 + 2 = _____

20) 4 + 2 = _____

21) 8 + 9 = _____

22) 8 + 7 = _____

23) 0 + 13 = _____

24) 2 + 17 = _____

25) 13 + 3 = _____

26) 1 + 17 = _____

27) 6 + 5 = _____

28) 11 + 3 = _____

29) 6 + 11 = _____

30) 4 + 10 = _____

31) 7 + 8 = _____

32) 18 + 2 = _____

33) 8 + 10 = _____

34) 9 + 5 = _____

35) 1 + 10 = _____

36) 4 + 6 = _____

37) 2 + 3 = _____

38) 2 + 15 = _____

39) 18 + 2 = _____

40) 5 + 6 = _____

41) 2 + 15 = _____

42) 13 + 4 = _____

43) 5 + 16 = _____

44) 5 + 15 = _____

45) 6 + 8 = _____

46) 14 + 6 = _____

47) 3 + 9 = _____

48) 11 + 7 = _____

49) 3 + 9 = _____

50) 13 + 3 = _____

51) 6 + 2 = _____

52) 17 + 3 = _____

53) 1 + 15 = _____

54) 5 + 4 = _____

55) 12 + 6 = _____

56) 3 + 16 = _____

57) 2 + 12 = _____

58) 10 + 5 = _____

59) 3 + 17 = _____

60) 17 + 1 = _____

Answer Key
Addition Worksheet 0 - 20

1) 1 + 17 = **18**

2) 5 + 13 = **18**

3) 3 + 8 = **11**

4) 17 + 1 = **18**

5) 20 + 0 = **20**

6) 5 + 10 = **15**

7) 8 + 6 = **14**

8) 12 + 6 = **18**

9) 3 + 11 = **14**

10) 4 + 16 = **20**

11) 10 + 9 = **19**

12) 12 + 2 = **14**

13) 3 + 14 = **17**

14) 4 + 12 = **16**

15) 13 + 5 = **18**

16) 3 + 14 = **17**

17) 14 + 2 = **16**

18) 5 + 2 = **7**

19) 18 + 2 = **20**

20) 4 + 2 = **6**

21) 8 + 9 = **17**

22) 8 + 7 = **15**

23) 0 + 13 = **13**

24) 2 + 17 = **19**

25) 13 + 3 = **16**

26) 1 + 17 = **18**

27) 6 + 5 = **11**

28) 11 + 3 = **14**

29) 6 + 11 = **17**

30) 4 + 10 = **14**

31) 7 + 8 = **15**

32) 18 + 2 = **20**

33) 8 + 10 = **18**

34) 9 + 5 = **14**

35) 1 + 10 = **11**

36) 4 + 6 = **10**

37) 2 + 3 = **5**

38) 2 + 15 = **17**

39) 18 + 2 = **20**

40) 5 + 6 = **11**

41) 2 + 15 = **17**

42) 13 + 4 = **17**

43) 5 + 16 = **21**

44) 5 + 15 = **20**

45) 6 + 8 = **14**

46) 14 + 6 = **20**

47) 3 + 9 = **12**

48) 11 + 7 = **18**

49) 3 + 9 = **12**

50) 13 + 3 = **16**

51) 6 + 2 = **8**

52) 17 + 3 = **20**

53) 1 + 15 = **16**

54) 5 + 4 = **9**

55) 12 + 6 = **18**

56) 3 + 16 = **19**

57) 2 + 12 = **14**

58) 10 + 5 = **15**

59) 3 + 17 = **20**

60) 17 + 1 = **18**

Addition Worksheet 0 - 20

1) 4 + 16 = _____   2) 4 + 10 = _____   3) 5 + 1 = _____

4) 15 + 1 = _____   5) 0 + 20 = _____   6) 6 + 10 = _____

7) 5 + 3 = _____   8) 8 + 5 = _____   9) 13 + 5 = _____

10) 1 + 18 = _____   11) 2 + 16 = _____   12) 5 + 3 = _____

13) 12 + 7 = _____   14) 2 + 18 = _____   15) 15 + 5 = _____

16) 5 + 2 = _____   17) 14 + 1 = _____   18) 4 + 15 = _____

19) 6 + 5 = _____   20) 11 + 8 = _____   21) 3 + 14 = _____

22) 16 + 3 = _____   23) 1 + 10 = _____   24) 3 + 7 = _____

25) 11 + 5 = _____   26) 2 + 17 = _____   27) 1 + 13 = _____

28) 4 + 12 = _____   29) 17 + 2 = _____   30) 11 + 3 = _____

31) 3 + 15 = _____   32) 9 + 5 = _____   33) 11 + 7 = _____

34) 16 + 3 = _____   35) 2 + 8 = _____   36) 5 + 12 = _____

37) 11 + 0 = _____   38) 4 + 5 = _____   39) 7 + 1 = _____

40) 1 + 15 = _____   41) 5 + 15 = _____   42) 15 + 3 = _____

43) 3 + 14 = _____   44) 17 + 2 = _____   45) 9 + 5 = _____

46) 12 + 4 = _____   47) 6 + 7 = _____   48) 12 + 3 = _____

49) 2 + 8 = _____   50) 2 + 18 = _____   51) 0 + 20 = _____

52) 9 + 7 = _____   53) 4 + 14 = _____   54) 1 + 19 = _____

55) 3 + 12 = _____   56) 2 + 2 = _____   57) 2 + 9 = _____

58) 5 + 12 = _____   59) 20 + 0 = _____   60) 6 + 5 = _____

Answer Key
Addition Worksheet 0 - 20

1) 4 + 16 = **20**   2) 4 + 10 = **14**   3) 5 + 1 = **6**

4) 15 + 1 = **16**   5) 0 + 20 = **20**   6) 6 + 10 = **16**

7) 5 + 3 = **8**   8) 8 + 5 = **13**   9) 13 + 5 = **18**

10) 1 + 18 = **19**   11) 2 + 16 = **18**   12) 5 + 3 = **8**

13) 12 + 7 = **19**   14) 2 + 18 = **20**   15) 15 + 5 = **20**

16) 5 + 2 = **7**   17) 14 + 1 = **15**   18) 4 + 15 = **19**

19) 6 + 5 = **11**   20) 11 + 8 = **19**   21) 3 + 14 = **17**

22) 16 + 3 = **19**   23) 1 + 10 = **11**   24) 3 + 7 = **10**

25) 11 + 5 = **16**   26) 2 + 17 = **19**   27) 1 + 13 = **14**

28) 4 + 12 = **16**   29) 17 + 2 = **19**   30) 11 + 3 = **14**

31) 3 + 15 = **18**   32) 9 + 5 = **14**   33) 11 + 7 = **18**

34) 16 + 3 = **19**   35) 2 + 8 = **10**   36) 5 + 12 = **17**

37) 11 + 0 = **11**   38) 4 + 5 = **9**   39) 7 + 1 = **8**

40) 1 + 15 = **16**   41) 5 + 15 = **20**   42) 15 + 3 = **18**

43) 3 + 14 = **17**   44) 17 + 2 = **19**   45) 9 + 5 = **14**

46) 12 + 4 = **16**   47) 6 + 7 = **13**   48) 12 + 3 = **15**

49) 2 + 8 = **10**   50) 2 + 18 = **20**   51) 0 + 20 = **20**

52) 9 + 7 = **16**   53) 4 + 14 = **18**   54) 1 + 19 = **20**

55) 3 + 12 = **15**   56) 2 + 2 = **4**   57) 2 + 9 = **11**

58) 5 + 12 = **17**   59) 20 + 0 = **20**   60) 6 + 5 = **11**

Date : _________

Time : _________

Name : __________

Score : ______/60

Addition Worksheet 0 - 20

1) 1 + 18 = ______

2) 12 + 4 = ______

3) 8 + 3 = ______

4) 11 + 9 = ______

5) 9 + 8 = ______

6) 6 + 7 = ______

7) 8 + 12 = ______

8) 7 + 1 = ______

9) 13 + 7 = ______

10) 13 + 2 = ______

11) 1 + 8 = ______

12) 4 + 10 = ______

13) 3 + 11 = ______

14) 2 + 1 = ______

15) 6 + 10 = ______

16) 1 + 6 = ______

17) 2 + 9 = ______

18) 11 + 4 = ______

19) 19 + 1 = ______

20) 1 + 10 = ______

21) 0 + 20 = ______

22) 3 + 17 = ______

23) 3 + 11 = ______

24) 4 + 13 = ______

25) 13 + 4 = ______

26) 16 + 1 = ______

27) 2 + 16 = ______

28) 4 + 13 = ______

29) 10 + 7 = ______

30) 20 + 0 = ______

31) 15 + 5 = ______

32) 19 + 1 = ______

33) 20 + 0 = ______

34) 16 + 1 = ______

35) 6 + 13 = ______

36) 16 + 1 = ______

37) 10 + 2 = ______

38) 0 + 10 = ______

39) 1 + 13 = ______

40) 10 + 5 = ______

41) 8 + 1 = ______

42) 5 + 14 = ______

43) 16 + 2 = ______

44) 13 + 6 = ______

45) 18 + 1 = ______

46) 1 + 18 = ______

47) 4 + 9 = ______

48) 13 + 1 = ______

49) 2 + 18 = ______

50) 15 + 5 = ______

51) 5 + 4 = ______

52) 2 + 15 = ______

53) 12 + 5 = ______

54) 3 + 1 = ______

55) 0 + 9 = ______

56) 11 + 7 = ______

57) 1 + 11 = ______

58) 8 + 1 = ______

59) 10 + 7 = ______

60) 1 + 18 = ______

Answer Key
Addition Worksheet 0 - 20

1) $1 + 18 = \mathbf{19}$ 2) $12 + 4 = \mathbf{16}$ 3) $8 + 3 = \mathbf{11}$

4) $11 + 9 = \mathbf{20}$ 5) $9 + 8 = \mathbf{17}$ 6) $6 + 7 = \mathbf{13}$

7) $8 + 12 = \mathbf{20}$ 8) $7 + 1 = \mathbf{8}$ 9) $13 + 7 = \mathbf{20}$

10) $13 + 2 = \mathbf{15}$ 11) $1 + 8 = \mathbf{9}$ 12) $4 + 10 = \mathbf{14}$

13) $3 + 11 = \mathbf{14}$ 14) $2 + 1 = \mathbf{3}$ 15) $6 + 10 = \mathbf{16}$

16) $1 + 6 = \mathbf{7}$ 17) $2 + 9 = \mathbf{11}$ 18) $11 + 4 = \mathbf{15}$

19) $19 + 1 = \mathbf{20}$ 20) $1 + 10 = \mathbf{11}$ 21) $0 + 20 = \mathbf{20}$

22) $3 + 17 = \mathbf{20}$ 23) $3 + 11 = \mathbf{14}$ 24) $4 + 13 = \mathbf{17}$

25) $13 + 4 = \mathbf{17}$ 26) $16 + 1 = \mathbf{17}$ 27) $2 + 16 = \mathbf{18}$

28) $4 + 13 = \mathbf{17}$ 29) $10 + 7 = \mathbf{17}$ 30) $20 + 0 = \mathbf{20}$

31) $15 + 5 = \mathbf{20}$ 32) $19 + 1 = \mathbf{20}$ 33) $20 + 0 = \mathbf{20}$

34) $16 + 1 = \mathbf{17}$ 35) $6 + 13 = \mathbf{19}$ 36) $16 + 1 = \mathbf{17}$

37) $10 + 2 = \mathbf{12}$ 38) $0 + 10 = \mathbf{10}$ 39) $1 + 13 = \mathbf{14}$

40) $10 + 5 = \mathbf{15}$ 41) $8 + 1 = \mathbf{9}$ 42) $5 + 14 = \mathbf{19}$

43) $16 + 2 = \mathbf{18}$ 44) $13 + 6 = \mathbf{19}$ 45) $18 + 1 = \mathbf{19}$

46) $1 + 18 = \mathbf{19}$ 47) $4 + 9 = \mathbf{13}$ 48) $13 + 1 = \mathbf{14}$

49) $2 + 18 = \mathbf{20}$ 50) $15 + 5 = \mathbf{20}$ 51) $5 + 4 = \mathbf{9}$

52) $2 + 15 = \mathbf{17}$ 53) $12 + 5 = \mathbf{17}$ 54) $3 + 1 = \mathbf{4}$

55) $0 + 9 = \mathbf{9}$ 56) $11 + 7 = \mathbf{18}$ 57) $1 + 11 = \mathbf{12}$

58) $8 + 1 = \mathbf{9}$ 59) $10 + 7 = \mathbf{17}$ 60) $1 + 18 = \mathbf{19}$

Addition Worksheet 0 - 20

1) 2 + 9 = _____	2) 2 + 10 = _____	3) 10 + 1 = _____
4) 10 + 1 = _____	5) 13 + 5 = _____	6) 7 + 5 = _____
7) 5 + 5 = _____	8) 6 + 5 = _____	9) 10 + 8 = _____
10) 1 + 10 = _____	11) 5 + 5 = _____	12) 19 + 1 = _____
13) 1 + 19 = _____	14) 5 + 13 = _____	15) 11 + 7 = _____
16) 9 + 7 = _____	17) 0 + 20 = _____	18) 6 + 12 = _____
19) 8 + 6 = _____	20) 1 + 16 = _____	21) 6 + 5 = _____
22) 6 + 7 = _____	23) 3 + 5 = _____	24) 18 + 1 = _____
25) 7 + 8 = _____	26) 2 + 18 = _____	27) 17 + 1 = _____
28) 9 + 1 = _____	29) 20 + 0 = _____	30) 3 + 12 = _____
31) 17 + 3 = _____	32) 4 + 14 = _____	33) 2 + 18 = _____
34) 4 + 6 = _____	35) 4 + 9 = _____	36) 3 + 9 = _____
37) 9 + 10 = _____	38) 18 + 1 = _____	39) 1 + 4 = _____
40) 9 + 8 = _____	41) 7 + 7 = _____	42) 6 + 7 = _____
43) 1 + 17 = _____	44) 4 + 0 = _____	45) 9 + 6 = _____
46) 5 + 14 = _____	47) 2 + 17 = _____	48) 3 + 1 = _____
49) 9 + 7 = _____	50) 9 + 3 = _____	51) 17 + 2 = _____
52) 8 + 1 = _____	53) 15 + 3 = _____	54) 3 + 13 = _____
55) 5 + 4 = _____	56) 4 + 10 = _____	57) 5 + 3 = _____
58) 1 + 11 = _____	59) 16 + 3 = _____	60) 2 + 12 = _____

Answer Key
Addition Worksheet 0 - 20

1) 2 + 9 = **11**

2) 2 + 10 = **12**

3) 10 + 1 = **11**

4) 10 + 1 = **11**

5) 13 + 5 = **18**

6) 7 + 5 = **12**

7) 5 + 5 = **10**

8) 6 + 5 = **11**

9) 10 + 8 = **18**

10) 1 + 10 = **11**

11) 5 + 5 = **10**

12) 19 + 1 = **20**

13) 1 + 19 = **20**

14) 5 + 13 = **18**

15) 11 + 7 = **18**

16) 9 + 7 = **16**

17) 0 + 20 = **20**

18) 6 + 12 = **18**

19) 8 + 6 = **14**

20) 1 + 16 = **17**

21) 6 + 5 = **11**

22) 6 + 7 = **13**

23) 3 + 5 = **8**

24) 18 + 1 = **19**

25) 7 + 8 = **15**

26) 2 + 18 = **20**

27) 17 + 1 = **18**

28) 9 + 1 = **10**

29) 20 + 0 = **20**

30) 3 + 12 = **15**

31) 17 + 3 = **20**

32) 4 + 14 = **18**

33) 2 + 18 = **20**

34) 4 + 6 = **10**

35) 4 + 9 = **13**

36) 3 + 9 = **12**

37) 9 + 10 = **19**

38) 18 + 1 = **19**

39) 1 + 4 = **5**

40) 9 + 8 = **17**

41) 7 + 7 = **14**

42) 6 + 7 = **13**

43) 1 + 17 = **18**

44) 4 + 0 = **4**

45) 9 + 6 = **15**

46) 5 + 14 = **19**

47) 2 + 17 = **19**

48) 3 + 1 = **4**

49) 9 + 7 = **16**

50) 9 + 3 = **12**

51) 17 + 2 = **19**

52) 8 + 1 = **9**

53) 15 + 3 = **18**

54) 3 + 13 = **16**

55) 5 + 4 = **9**

56) 4 + 10 = **14**

57) 5 + 3 = **8**

58) 1 + 11 = **12**

59) 16 + 3 = **19**

60) 2 + 12 = **14**

Date : _________   Name : _________
Time : _________   Score : _____/60

Addition Worksheet 0 - 20

1) 2 + 11 = _____   2) 2 + 18 = _____   3) 1 + 16 = _____

4) 14 + 2 = _____   5) 10 + 10 = _____   6) 4 + 7 = _____

7) 1 + 19 = _____   8) 12 + 4 = _____   9) 5 + 7 = _____

10) 19 + 1 = _____   11) 14 + 5 = _____   12) 9 + 2 = _____

13) 1 + 13 = _____   14) 2 + 3 = _____   15) 3 + 10 = _____

16) 13 + 5 = _____   17) 1 + 11 = _____   18) 2 + 16 = _____

19) 13 + 2 = _____   20) 5 + 14 = _____   21) 1 + 2 = _____

22) 15 + 4 = _____   23) 11 + 3 = _____   24) 11 + 7 = _____

25) 0 + 20 = _____   26) 12 + 2 = _____   27) 4 + 6 = _____

28) 16 + 3 = _____   29) 7 + 10 = _____   30) 1 + 13 = _____

31) 4 + 7 = _____   32) 3 + 15 = _____   33) 14 + 2 = _____

34) 4 + 14 = _____   35) 3 + 2 = _____   36) 8 + 3 = _____

37) 1 + 19 = _____   38) 8 + 12 = _____   39) 4 + 11 = _____

40) 19 + 1 = _____   41) 2 + 9 = _____   42) 15 + 5 = _____

43) 19 + 1 = _____   44) 12 + 6 = _____   45) 3 + 13 = _____

46) 15 + 3 = _____   47) 14 + 4 = _____   48) 17 + 1 = _____

49) 14 + 6 = _____   50) 9 + 3 = _____   51) 10 + 9 = _____

52) 18 + 0 = _____   53) 4 + 10 = _____   54) 1 + 5 = _____

55) 4 + 16 = _____   56) 2 + 10 = _____   57) 2 + 8 = _____

58) 12 + 5 = _____   59) 15 + 4 = _____   60) 20 + 0 = _____

Answer Key
Addition Worksheet 0 - 20

1) $2 + 11 = \mathbf{13}$

2) $2 + 18 = \mathbf{20}$

3) $1 + 16 = \mathbf{17}$

4) $14 + 2 = \mathbf{16}$

5) $10 + 10 = \mathbf{20}$

6) $4 + 7 = \mathbf{11}$

7) $1 + 19 = \mathbf{20}$

8) $12 + 4 = \mathbf{16}$

9) $5 + 7 = \mathbf{12}$

10) $19 + 1 = \mathbf{20}$

11) $14 + 5 = \mathbf{19}$

12) $9 + 2 = \mathbf{11}$

13) $1 + 13 = \mathbf{14}$

14) $2 + 3 = \mathbf{5}$

15) $3 + 10 = \mathbf{13}$

16) $13 + 5 = \mathbf{18}$

17) $1 + 11 = \mathbf{12}$

18) $2 + 16 = \mathbf{18}$

19) $13 + 2 = \mathbf{15}$

20) $5 + 14 = \mathbf{19}$

21) $1 + 2 = \mathbf{3}$

22) $15 + 4 = \mathbf{19}$

23) $11 + 3 = \mathbf{14}$

24) $11 + 7 = \mathbf{18}$

25) $0 + 20 = \mathbf{20}$

26) $12 + 2 = \mathbf{14}$

27) $4 + 6 = \mathbf{10}$

28) $16 + 3 = \mathbf{19}$

29) $7 + 10 = \mathbf{17}$

30) $1 + 13 = \mathbf{14}$

31) $4 + 7 = \mathbf{11}$

32) $3 + 15 = \mathbf{18}$

33) $14 + 2 = \mathbf{16}$

34) $4 + 14 = \mathbf{18}$

35) $3 + 2 = \mathbf{5}$

36) $8 + 3 = \mathbf{11}$

37) $1 + 19 = \mathbf{20}$

38) $8 + 12 = \mathbf{20}$

39) $4 + 11 = \mathbf{15}$

40) $19 + 1 = \mathbf{20}$

41) $2 + 9 = \mathbf{11}$

42) $15 + 5 = \mathbf{20}$

43) $19 + 1 = \mathbf{20}$

44) $12 + 6 = \mathbf{18}$

45) $3 + 13 = \mathbf{16}$

46) $15 + 3 = \mathbf{18}$

47) $14 + 4 = \mathbf{18}$

48) $17 + 1 = \mathbf{18}$

49) $14 + 6 = \mathbf{20}$

50) $9 + 3 = \mathbf{12}$

51) $10 + 9 = \mathbf{19}$

52) $18 + 0 = \mathbf{18}$

53) $4 + 10 = \mathbf{14}$

54) $1 + 5 = \mathbf{6}$

55) $4 + 16 = \mathbf{20}$

56) $2 + 10 = \mathbf{12}$

57) $2 + 8 = \mathbf{10}$

58) $12 + 5 = \mathbf{17}$

59) $15 + 4 = \mathbf{19}$

60) $20 + 0 = \mathbf{20}$

Addition Worksheet 0 - 20

1) 2 + 18 = _____	2) 1 + 19 = _____	3) 10 + 10 = _____
4) 4 + 7 = _____	5) 18 + 1 = _____	6) 3 + 8 = _____
7) 7 + 13 = _____	8) 8 + 8 = _____	9) 7 + 3 = _____
10) 2 + 16 = _____	11) 4 + 10 = _____	12) 12 + 4 = _____
13) 8 + 9 = _____	14) 13 + 6 = _____	15) 7 + 9 = _____
16) 0 + 18 = _____	17) 2 + 1 = _____	18) 7 + 10 = _____
19) 10 + 2 = _____	20) 5 + 9 = _____	21) 3 + 11 = _____
22) 7 + 11 = _____	23) 5 + 5 = _____	24) 1 + 16 = _____
25) 20 + 0 = _____	26) 3 + 7 = _____	27) 14 + 4 = _____
28) 7 + 6 = _____	29) 20 + 0 = _____	30) 8 + 8 = _____
31) 2 + 18 = _____	32) 4 + 14 = _____	33) 10 + 5 = _____
34) 3 + 15 = _____	35) 4 + 7 = _____	36) 1 + 18 = _____
37) 1 + 18 = _____	38) 5 + 14 = _____	39) 4 + 4 = _____
40) 4 + 10 = _____	41) 1 + 1 = _____	42) 8 + 11 = _____
43) 2 + 1 = _____	44) 19 + 1 = _____	45) 11 + 5 = _____
46) 13 + 4 = _____	47) 3 + 4 = _____	48) 16 + 4 = _____
49) 10 + 8 = _____	50) 2 + 13 = _____	51) 7 + 8 = _____
52) 8 + 8 = _____	53) 8 + 10 = _____	54) 3 + 13 = _____
55) 12 + 4 = _____	56) 14 + 2 = _____	57) 12 + 5 = _____
58) 16 + 4 = _____	59) 4 + 1 = _____	60) 10 + 8 = _____

Answer Key
Addition Worksheet 0 - 20

1) 2 + 18 = **20**

2) 1 + 19 = **20**

3) 10 + 10 = **20**

4) 4 + 7 = **11**

5) 18 + 1 = **19**

6) 3 + 8 = **11**

7) 7 + 13 = **20**

8) 8 + 8 = **16**

9) 7 + 3 = **10**

10) 2 + 16 = **18**

11) 4 + 10 = **14**

12) 12 + 4 = **16**

13) 8 + 9 = **17**

14) 13 + 6 = **19**

15) 7 + 9 = **16**

16) 0 + 18 = **18**

17) 2 + 1 = **3**

18) 7 + 10 = **17**

19) 10 + 2 = **12**

20) 5 + 9 = **14**

21) 3 + 11 = **14**

22) 7 + 11 = **18**

23) 5 + 5 = **10**

24) 1 + 16 = **17**

25) 20 + 0 = **20**

26) 3 + 7 = **10**

27) 14 + 4 = **18**

28) 7 + 6 = **13**

29) 20 + 0 = **20**

30) 8 + 8 = **16**

31) 2 + 18 = **20**

32) 4 + 14 = **18**

33) 10 + 5 = **15**

34) 3 + 15 = **18**

35) 4 + 7 = **11**

36) 1 + 18 = **19**

37) 1 + 18 = **19**

38) 5 + 14 = **19**

39) 4 + 4 = **8**

40) 4 + 10 = **14**

41) 1 + 1 = **2**

42) 8 + 11 = **19**

43) 2 + 1 = **3**

44) 19 + 1 = **20**

45) 11 + 5 = **16**

46) 13 + 4 = **17**

47) 3 + 4 = **7**

48) 16 + 4 = **20**

49) 10 + 8 = **18**

50) 2 + 13 = **15**

51) 7 + 8 = **15**

52) 8 + 8 = **16**

53) 8 + 10 = **18**

54) 3 + 13 = **16**

55) 12 + 4 = **16**

56) 14 + 2 = **16**

57) 12 + 5 = **17**

58) 16 + 4 = **20**

59) 4 + 1 = **5**

60) 10 + 8 = **18**

Date : _________

Time : _________

Name : _________

Score : _____/60

Addition Worksheet 0 - 20

1) 15 + 5 = _____

2) 2 + 18 = _____

3) 8 + 4 = _____

4) 2 + 17 = _____

5) 2 + 5 = _____

6) 14 + 2 = _____

7) 7 + 13 = _____

8) 15 + 3 = _____

9) 5 + 15 = _____

10) 9 + 2 = _____

11) 4 + 4 = _____

12) 2 + 5 = _____

13) 1 + 8 = _____

14) 4 + 6 = _____

15) 15 + 3 = _____

16) 4 + 15 = _____

17) 10 + 2 = _____

18) 7 + 8 = _____

19) 6 + 6 = _____

20) 18 + 2 = _____

21) 2 + 13 = _____

22) 5 + 3 = _____

23) 18 + 1 = _____

24) 7 + 13 = _____

25) 3 + 5 = _____

26) 2 + 9 = _____

27) 3 + 16 = _____

28) 1 + 15 = _____

29) 18 + 2 = _____

30) 12 + 1 = _____

31) 3 + 16 = _____

32) 8 + 11 = _____

33) 5 + 9 = _____

34) 3 + 13 = _____

35) 17 + 2 = _____

36) 11 + 2 = _____

37) 5 + 10 = _____

38) 1 + 18 = _____

39) 15 + 3 = _____

40) 1 + 17 = _____

41) 9 + 9 = _____

42) 9 + 4 = _____

43) 5 + 10 = _____

44) 18 + 1 = _____

45) 4 + 11 = _____

46) 2 + 4 = _____

47) 0 + 5 = _____

48) 7 + 7 = _____

49) 2 + 17 = _____

50) 3 + 11 = _____

51) 3 + 2 = _____

52) 2 + 3 = _____

53) 8 + 7 = _____

54) 16 + 2 = _____

55) 13 + 5 = _____

56) 17 + 3 = _____

57) 1 + 18 = _____

58) 8 + 10 = _____

59) 1 + 9 = _____

60) 13 + 5 = _____

Answer Key
Addition Worksheet 0 - 20

1) 15 + 5 = **20**

2) 2 + 18 = **20**

3) 8 + 4 = **12**

4) 2 + 17 = **19**

5) 2 + 5 = **7**

6) 14 + 2 = **16**

7) 7 + 13 = **20**

8) 15 + 3 = **18**

9) 5 + 15 = **20**

10) 9 + 2 = **11**

11) 4 + 4 = **8**

12) 2 + 5 = **7**

13) 1 + 8 = **9**

14) 4 + 6 = **10**

15) 15 + 3 = **18**

16) 4 + 15 = **19**

17) 10 + 2 = **12**

18) 7 + 8 = **15**

19) 6 + 6 = **12**

20) 18 + 2 = **20**

21) 2 + 13 = **15**

22) 5 + 3 = **8**

23) 18 + 1 = **19**

24) 7 + 13 = **20**

25) 3 + 5 = **8**

26) 2 + 9 = **11**

27) 3 + 16 = **19**

28) 1 + 15 = **16**

29) 18 + 2 = **20**

30) 12 + 1 = **13**

31) 3 + 16 = **19**

32) 8 + 11 = **19**

33) 5 + 9 = **14**

34) 3 + 13 = **16**

35) 17 + 2 = **19**

36) 11 + 2 = **13**

37) 5 + 10 = **15**

38) 1 + 18 = **19**

39) 15 + 3 = **18**

40) 1 + 17 = **18**

41) 9 + 9 = **18**

42) 9 + 4 = **13**

43) 5 + 10 = **15**

44) 18 + 1 = **19**

45) 4 + 11 = **15**

46) 2 + 4 = **6**

47) 0 + 5 = **5**

48) 7 + 7 = **14**

49) 2 + 17 = **19**

50) 3 + 11 = **14**

51) 3 + 2 = **5**

52) 2 + 3 = **5**

53) 8 + 7 = **15**

54) 16 + 2 = **18**

55) 13 + 5 = **18**

56) 17 + 3 = **20**

57) 1 + 18 = **19**

58) 8 + 10 = **18**

59) 1 + 9 = **10**

60) 13 + 5 = **18**

Date : _________   Name : __________

Time : _________   Score : _____/60

Addition Worksheet 0 - 20

1) 17 + 3 = _____   2) 2 + 14 = _____   3) 1 + 3 = _____

4) 7 + 1 = _____   5) 10 + 2 = _____   6) 7 + 4 = _____

7) 5 + 12 = _____   8) 0 + 20 = _____   9) 5 + 10 = _____

10) 2 + 18 = _____   11) 1 + 14 = _____   12) 2 + 7 = _____

13) 13 + 6 = _____   14) 20 + 0 = _____   15) 10 + 1 = _____

16) 11 + 8 = _____   17) 3 + 10 = _____   18) 17 + 3 = _____

19) 18 + 0 = _____   20) 5 + 13 = _____   21) 15 + 3 = _____

22) 0 + 20 = _____   23) 2 + 18 = _____   24) 12 + 4 = _____

25) 6 + 11 = _____   26) 12 + 7 = _____   27) 6 + 7 = _____

28) 0 + 4 = _____   29) 1 + 6 = _____   30) 1 + 17 = _____

31) 4 + 15 = _____   32) 11 + 7 = _____   33) 6 + 3 = _____

34) 15 + 1 = _____   35) 2 + 14 = _____   36) 1 + 17 = _____

37) 1 + 19 = _____   38) 5 + 13 = _____   39) 6 + 12 = _____

40) 8 + 3 = _____   41) 5 + 12 = _____   42) 19 + 0 = _____

43) 1 + 19 = _____   44) 8 + 6 = _____   45) 4 + 4 = _____

46) 17 + 2 = _____   47) 4 + 13 = _____   48) 11 + 8 = _____

49) 1 + 18 = _____   50) 4 + 16 = _____   51) 15 + 3 = _____

52) 1 + 6 = _____   53) 1 + 19 = _____   54) 13 + 1 = _____

55) 4 + 2 = _____   56) 2 + 13 = _____   57) 2 + 12 = _____

58) 6 + 4 = _____   59) 5 + 15 = _____   60) 1 + 18 = _____

Answer Key

Addition Worksheet 0 - 20

1) 17 + 3 = **20**

2) 2 + 14 = **16**

3) 1 + 3 = **4**

4) 7 + 1 = **8**

5) 10 + 2 = **12**

6) 7 + 4 = **11**

7) 5 + 12 = **17**

8) 0 + 20 = **20**

9) 5 + 10 = **15**

10) 2 + 18 = **20**

11) 1 + 14 = **15**

12) 2 + 7 = **9**

13) 13 + 6 = **19**

14) 20 + 0 = **20**

15) 10 + 1 = **11**

16) 11 + 8 = **19**

17) 3 + 10 = **13**

18) 17 + 3 = **20**

19) 18 + 0 = **18**

20) 5 + 13 = **18**

21) 15 + 3 = **18**

22) 0 + 20 = **20**

23) 2 + 18 = **20**

24) 12 + 4 = **16**

25) 6 + 11 = **17**

26) 12 + 7 = **19**

27) 6 + 7 = **13**

28) 0 + 4 = **4**

29) 1 + 6 = **7**

30) 1 + 17 = **18**

31) 4 + 15 = **19**

32) 11 + 7 = **18**

33) 6 + 3 = **9**

34) 15 + 1 = **16**

35) 2 + 14 = **16**

36) 1 + 17 = **18**

37) 1 + 19 = **20**

38) 5 + 13 = **18**

39) 6 + 12 = **18**

40) 8 + 3 = **11**

41) 5 + 12 = **17**

42) 19 + 0 = **19**

43) 1 + 19 = **20**

44) 8 + 6 = **14**

45) 4 + 4 = **8**

46) 17 + 2 = **19**

47) 4 + 13 = **17**

48) 11 + 8 = **19**

49) 1 + 18 = **19**

50) 4 + 16 = **20**

51) 15 + 3 = **18**

52) 1 + 6 = **7**

53) 1 + 19 = **20**

54) 13 + 1 = **14**

55) 4 + 2 = **6**

56) 2 + 13 = **15**

57) 2 + 12 = **14**

58) 6 + 4 = **10**

59) 5 + 15 = **20**

60) 1 + 18 = **19**

Addition Worksheet 0 - 20

1) 14 + 1 = _____	2) 7 + 11 = _____	3) 3 + 1 = _____
4) 8 + 1 = _____	5) 5 + 11 = _____	6) 1 + 15 = _____
7) 17 + 0 = _____	8) 1 + 18 = _____	9) 13 + 6 = _____
10) 4 + 10 = _____	11) 3 + 8 = _____	12) 8 + 7 = _____
13) 2 + 10 = _____	14) 18 + 2 = _____	15) 13 + 2 = _____
16) 3 + 6 = _____	17) 0 + 20 = _____	18) 4 + 16 = _____
19) 14 + 3 = _____	20) 0 + 20 = _____	21) 4 + 14 = _____
22) 13 + 6 = _____	23) 0 + 12 = _____	24) 14 + 3 = _____
25) 15 + 4 = _____	26) 2 + 6 = _____	27) 15 + 5 = _____
28) 14 + 4 = _____	29) 5 + 11 = _____	30) 20 + 0 = _____
31) 1 + 16 = _____	32) 1 + 18 = _____	33) 9 + 9 = _____
34) 9 + 5 = _____	35) 3 + 13 = _____	36) 7 + 13 = _____
37) 14 + 6 = _____	38) 17 + 2 = _____	39) 1 + 8 = _____
40) 4 + 7 = _____	41) 12 + 3 = _____	42) 3 + 17 = _____
43) 4 + 14 = _____	44) 15 + 4 = _____	45) 0 + 5 = _____
46) 1 + 16 = _____	47) 7 + 10 = _____	48) 7 + 4 = _____
49) 1 + 3 = _____	50) 9 + 1 = _____	51) 3 + 16 = _____
52) 10 + 5 = _____	53) 2 + 14 = _____	54) 17 + 2 = _____
55) 10 + 2 = _____	56) 1 + 18 = _____	57) 4 + 15 = _____
58) 2 + 3 = _____	59) 4 + 13 = _____	60) 12 + 6 = _____

Answer Key

Addition Worksheet 0 - 20

1) 14 + 1 = **15**

2) 7 + 11 = **18**

3) 3 + 1 = **4**

4) 8 + 1 = **9**

5) 5 + 11 = **16**

6) 1 + 15 = **16**

7) 17 + 0 = **17**

8) 1 + 18 = **19**

9) 13 + 6 = **19**

10) 4 + 10 = **14**

11) 3 + 8 = **11**

12) 8 + 7 = **15**

13) 2 + 10 = **12**

14) 18 + 2 = **20**

15) 13 + 2 = **15**

16) 3 + 6 = **9**

17) 0 + 20 = **20**

18) 4 + 16 = **20**

19) 14 + 3 = **17**

20) 0 + 20 = **20**

21) 4 + 14 = **18**

22) 13 + 6 = **19**

23) 0 + 12 = **12**

24) 14 + 3 = **17**

25) 15 + 4 = **19**

26) 2 + 6 = **8**

27) 15 + 5 = **20**

28) 14 + 4 = **18**

29) 5 + 11 = **16**

30) 20 + 0 = **20**

31) 1 + 16 = **17**

32) 1 + 18 = **19**

33) 9 + 9 = **18**

34) 9 + 5 = **14**

35) 3 + 13 = **16**

36) 7 + 13 = **20**

37) 14 + 6 = **20**

38) 17 + 2 = **19**

39) 1 + 8 = **9**

40) 4 + 7 = **11**

41) 12 + 3 = **15**

42) 3 + 17 = **20**

43) 4 + 14 = **18**

44) 15 + 4 = **19**

45) 0 + 5 = **5**

46) 1 + 16 = **17**

47) 7 + 10 = **17**

48) 7 + 4 = **11**

49) 1 + 3 = **4**

50) 9 + 1 = **10**

51) 3 + 16 = **19**

52) 10 + 5 = **15**

53) 2 + 14 = **16**

54) 17 + 2 = **19**

55) 10 + 2 = **12**

56) 1 + 18 = **19**

57) 4 + 15 = **19**

58) 2 + 3 = **5**

59) 4 + 13 = **17**

60) 12 + 6 = **18**

Addition Worksheet 0 - 20

1) 1 + 6 = _____	2) 12 + 6 = _____	3) 17 + 2 = _____
4) 16 + 2 = _____	5) 0 + 7 = _____	6) 6 + 4 = _____
7) 10 + 10 = _____	8) 6 + 11 = _____	9) 4 + 12 = _____
10) 1 + 18 = _____	11) 12 + 4 = _____	12) 3 + 4 = _____
13) 7 + 7 = _____	14) 19 + 1 = _____	15) 19 + 1 = _____
16) 7 + 8 = _____	17) 3 + 12 = _____	18) 8 + 3 = _____
19) 3 + 16 = _____	20) 12 + 8 = _____	21) 2 + 11 = _____
22) 1 + 9 = _____	23) 6 + 2 = _____	24) 2 + 15 = _____
25) 6 + 5 = _____	26) 9 + 4 = _____	27) 15 + 5 = _____
28) 19 + 1 = _____	29) 3 + 10 = _____	30) 17 + 2 = _____
31) 10 + 4 = _____	32) 8 + 2 = _____	33) 15 + 2 = _____
34) 12 + 7 = _____	35) 12 + 2 = _____	36) 8 + 6 = _____
37) 4 + 14 = _____	38) 4 + 14 = _____	39) 19 + 1 = _____
40) 2 + 18 = _____	41) 3 + 8 = _____	42) 1 + 18 = _____
43) 9 + 1 = _____	44) 1 + 10 = _____	45) 7 + 1 = _____
46) 15 + 2 = _____	47) 0 + 17 = _____	48) 13 + 3 = _____
49) 13 + 6 = _____	50) 2 + 1 = _____	51) 9 + 2 = _____
52) 1 + 17 = _____	53) 2 + 7 = _____	54) 2 + 13 = _____
55) 2 + 18 = _____	56) 3 + 2 = _____	57) 3 + 5 = _____
58) 4 + 15 = _____	59) 12 + 3 = _____	60) 15 + 1 = _____

Answer Key
Addition Worksheet 0 - 20

1) 1 + 6 = **7**

2) 12 + 6 = **18**

3) 17 + 2 = **19**

4) 16 + 2 = **18**

5) 0 + 7 = **7**

6) 6 + 4 = **10**

7) 10 + 10 = **20**

8) 6 + 11 = **17**

9) 4 + 12 = **16**

10) 1 + 18 = **19**

11) 12 + 4 = **16**

12) 3 + 4 = **7**

13) 7 + 7 = **14**

14) 19 + 1 = **20**

15) 19 + 1 = **20**

16) 7 + 8 = **15**

17) 3 + 12 = **15**

18) 8 + 3 = **11**

19) 3 + 16 = **19**

20) 12 + 8 = **20**

21) 2 + 11 = **13**

22) 1 + 9 = **10**

23) 6 + 2 = **8**

24) 2 + 15 = **17**

25) 6 + 5 = **11**

26) 9 + 4 = **13**

27) 15 + 5 = **20**

28) 19 + 1 = **20**

29) 3 + 10 = **13**

30) 17 + 2 = **19**

31) 10 + 4 = **14**

32) 8 + 2 = **10**

33) 15 + 2 = **17**

34) 12 + 7 = **19**

35) 12 + 2 = **14**

36) 8 + 6 = **14**

37) 4 + 14 = **18**

38) 4 + 14 = **18**

39) 19 + 1 = **20**

40) 2 + 18 = **20**

41) 3 + 8 = **11**

42) 1 + 18 = **19**

43) 9 + 1 = **10**

44) 1 + 10 = **11**

45) 7 + 1 = **8**

46) 15 + 2 = **17**

47) 0 + 17 = **17**

48) 13 + 3 = **16**

49) 13 + 6 = **19**

50) 2 + 1 = **3**

51) 9 + 2 = **11**

52) 1 + 17 = **18**

53) 2 + 7 = **9**

54) 2 + 13 = **15**

55) 2 + 18 = **20**

56) 3 + 2 = **5**

57) 3 + 5 = **8**

58) 4 + 15 = **19**

59) 12 + 3 = **15**

60) 15 + 1 = **16**

Subtraction Worksheet 0 - 10

1) 8 − 5 = _____
2) 9 − 6 = _____
3) 2 − 1 = _____

4) 7 − 6 = _____
5) 9 − 4 = _____
6) 7 − 3 = _____

7) 8 − 1 = _____
8) 7 − 4 = _____
9) 1 − 1 = _____

10) 10 − 6 = _____
11) 4 − 1 = _____
12) 8 − 8 = _____

13) 8 − 7 = _____
14) 2 − 1 = _____
15) 10 − 6 = _____

16) 9 − 5 = _____
17) 5 − 3 = _____
18) 9 − 5 = _____

19) 5 − 2 = _____
20) 10 − 7 = _____
21) 7 − 1 = _____

22) 6 − 1 = _____
23) 6 − 2 = _____
24) 9 − 2 = _____

25) 10 − 4 = _____
26) 3 − 3 = _____
27) 8 − 6 = _____

28) 7 − 6 = _____
29) 7 − 4 = _____
30) 7 − 5 = _____

31) 6 − 4 = _____
32) 10 − 8 = _____
33) 9 − 8 = _____

34) 5 − 4 = _____
35) 4 − 2 = _____
36) 6 − 6 = _____

37) 8 − 1 = _____
38) 4 − 3 = _____
39) 4 − 1 = _____

40) 7 − 5 = _____
41) 10 − 4 = _____
42) 9 − 4 = _____

43) 10 − 5 = _____
44) 5 − 3 = _____
45) 6 − 3 = _____

46) 5 − 1 = _____
47) 9 − 3 = _____
48) 9 − 3 = _____

49) 3 − 3 = _____
50) 3 − 1 = _____
51) 10 − 7 = _____

52) 7 − 3 = _____
53) 8 − 4 = _____
54) 10 − 2 = _____

55) 7 − 5 = _____
56) 8 − 3 = _____
57) 10 − 7 = _____

58) 6 − 1 = _____
59) 10 − 2 = _____
60) 10 − 6 = _____

Answer Key
Subtraction Worksheet 0 - 10

1) 8 – 5 = **3**   2) 9 – 6 = **3**   3) 2 – 1 = **1**

4) 7 – 6 = **1**   5) 9 – 4 = **5**   6) 7 – 3 = **4**

7) 8 – 1 = **7**   8) 7 – 4 = **3**   9) 1 – 1 = **0**

10) 10 – 6 = **4**   11) 4 – 1 = **3**   12) 8 – 8 = **0**

13) 8 – 7 = **1**   14) 2 – 1 = **1**   15) 10 – 6 = **4**

16) 9 – 5 = **4**   17) 5 – 3 = **2**   18) 9 – 5 = **4**

19) 5 – 2 = **3**   20) 10 – 7 = **3**   21) 7 – 1 = **6**

22) 6 – 1 = **5**   23) 6 – 2 = **4**   24) 9 – 2 = **7**

25) 10 – 4 = **6**   26) 3 – 3 = **0**   27) 8 – 6 = **2**

28) 7 – 6 = **1**   29) 7 – 4 = **3**   30) 7 – 5 = **2**

31) 6 – 4 = **2**   32) 10 – 8 = **2**   33) 9 – 8 = **1**

34) 5 – 4 = **1**   35) 4 – 2 = **2**   36) 6 – 6 = **0**

37) 8 – 1 = **7**   38) 4 – 3 = **1**   39) 4 – 1 = **3**

40) 7 – 5 = **2**   41) 10 – 4 = **6**   42) 9 – 4 = **5**

43) 10 – 5 = **5**   44) 5 – 3 = **2**   45) 6 – 3 = **3**

46) 5 – 1 = **4**   47) 9 – 3 = **6**   48) 9 – 3 = **6**

49) 3 – 3 = **0**   50) 3 – 1 = **2**   51) 10 – 7 = **3**

52) 7 – 3 = **4**   53) 8 – 4 = **4**   54) 10 – 2 = **8**

55) 7 – 5 = **2**   56) 8 – 3 = **5**   57) 10 – 7 = **3**

58) 6 – 1 = **5**   59) 10 – 2 = **8**   60) 10 – 6 = **4**

Date : _________

Time : _________

Name : __________

Score : _____/60

Subtraction Worksheet 0 - 10

1) 7 – 2 = _____   2) 7 – 3 = _____   3) 8 – 7 = _____

4) 2 – 1 = _____   5) 3 – 0 = _____   6) 7 – 6 = _____

7) 4 – 1 = _____   8) 9 – 9 = _____   9) 5 – 2 = _____

10) 9 – 0 = _____   11) 5 – 4 = _____   12) 4 – 4 = _____

13) 6 – 4 = _____   14) 10 – 3 = _____   15) 9 – 8 = _____

16) 9 – 4 = _____   17) 6 – 4 = _____   18) 9 – 9 = _____

19) 7 – 3 = _____   20) 7 – 4 = _____   21) 6 – 3 = _____

22) 10 – 7 = _____   23) 9 – 7 = _____   24) 9 – 2 = _____

25) 9 – 5 = _____   26) 8 – 1 = _____   27) 3 – 2 = _____

28) 10 – 1 = _____   29) 5 – 0 = _____   30) 4 – 4 = _____

31) 8 – 3 = _____   32) 6 – 1 = _____   33) 9 – 6 = _____

34) 10 – 2 = _____   35) 10 – 5 = _____   36) 7 – 3 = _____

37) 3 – 1 = _____   38) 8 – 1 = _____   39) 6 – 5 = _____

40) 6 – 3 = _____   41) 9 – 6 = _____   42) 8 – 7 = _____

43) 6 – 5 = _____   44) 5 – 3 = _____   45) 3 – 1 = _____

46) 10 – 7 = _____   47) 3 – 3 = _____   48) 5 – 2 = _____

49) 8 – 1 = _____   50) 9 – 9 = _____   51) 8 – 1 = _____

52) 9 – 5 = _____   53) 10 – 5 = _____   54) 10 – 2 = _____

55) 9 – 4 = _____   56) 9 – 3 = _____   57) 9 – 5 = _____

58) 9 – 6 = _____   59) 6 – 3 = _____   60) 8 – 6 = _____

Answer Key
Subtraction Worksheet 0 - 10

1) 7 – 2 = **5**	2) 7 – 3 = **4**	3) 8 – 7 = **1**
4) 2 – 1 = **1**	5) 3 – 0 = **3**	6) 7 – 6 = **1**
7) 4 – 1 = **3**	8) 9 – 9 = **0**	9) 5 – 2 = **3**
10) 9 – 0 = **9**	11) 5 – 4 = **1**	12) 4 – 4 = **0**
13) 6 – 4 = **2**	14) 10 – 3 = **7**	15) 9 – 8 = **1**
16) 9 – 4 = **5**	17) 6 – 4 = **2**	18) 9 – 9 = **0**
19) 7 – 3 = **4**	20) 7 – 4 = **3**	21) 6 – 3 = **3**
22) 10 – 7 = **3**	23) 9 – 7 = **2**	24) 9 – 2 = **7**
25) 9 – 5 = **4**	26) 8 – 1 = **7**	27) 3 – 2 = **1**
28) 10 – 1 = **9**	29) 5 – 0 = **5**	30) 4 – 4 = **0**
31) 8 – 3 = **5**	32) 6 – 1 = **5**	33) 9 – 6 = **3**
34) 10 – 2 = **8**	35) 10 – 5 = **5**	36) 7 – 3 = **4**
37) 3 – 1 = **2**	38) 8 – 1 = **7**	39) 6 – 5 = **1**
40) 6 – 3 = **3**	41) 9 – 6 = **3**	42) 8 – 7 = **1**
43) 6 – 5 = **1**	44) 5 – 3 = **2**	45) 3 – 1 = **2**
46) 10 – 7 = **3**	47) 3 – 3 = **0**	48) 5 – 2 = **3**
49) 8 – 1 = **7**	50) 9 – 9 = **0**	51) 8 – 1 = **7**
52) 9 – 5 = **4**	53) 10 – 5 = **5**	54) 10 – 2 = **8**
55) 9 – 4 = **5**	56) 9 – 3 = **6**	57) 9 – 5 = **4**
58) 9 – 6 = **3**	59) 6 – 3 = **3**	60) 8 – 6 = **2**

Date : _________

Time : ________

Name : __________

Score : ______/60

Subtraction Worksheet 0 - 10

1) 8 – 3 = _____

2) 9 – 3 = _____

3) 7 – 1 = _____

4) 8 – 1 = _____

5) 9 – 0 = _____

6) 9 – 6 = _____

7) 8 – 7 = _____

8) 9 – 2 = _____

9) 7 – 3 = _____

10) 6 – 1 = _____

11) 10 – 8 = _____

12) 10 – 9 = _____

13) 7 – 6 = _____

14) 10 – 7 = _____

15) 10 – 9 = _____

16) 8 – 1 = _____

17) 9 – 4 = _____

18) 6 – 6 = _____

19) 10 – 7 = _____

20) 7 – 6 = _____

21) 9 – 7 = _____

22) 5 – 1 = _____

23) 10 – 4 = _____

24) 3 – 2 = _____

25) 7 – 6 = _____

26) 6 – 2 = _____

27) 4 – 2 = _____

28) 10 – 9 = _____

29) 9 – 1 = _____

30) 2 – 1 = _____

31) 5 – 2 = _____

32) 8 – 4 = _____

33) 9 – 3 = _____

34) 9 – 3 = _____

35) 1 – 0 = _____

36) 7 – 6 = _____

37) 10 – 0 = _____

38) 8 – 5 = _____

39) 9 – 2 = _____

40) 3 – 2 = _____

41) 10 – 3 = _____

42) 9 – 7 = _____

43) 7 – 1 = _____

44) 3 – 1 = _____

45) 10 – 3 = _____

46) 9 – 1 = _____

47) 9 – 5 = _____

48) 10 – 6 = _____

49) 10 – 1 = _____

50) 7 – 6 = _____

51) 8 – 4 = _____

52) 9 – 7 = _____

53) 7 – 3 = _____

54) 3 – 3 = _____

55) 8 – 5 = _____

56) 8 – 3 = _____

57) 10 – 9 = _____

58) 10 – 3 = _____

59) 3 – 1 = _____

60) 3 – 2 = _____

Answer Key
Subtraction Worksheet 0 - 10

1) 8 − 3 = **5**

2) 9 − 3 = **6**

3) 7 − 1 = **6**

4) 8 − 1 = **7**

5) 9 − 0 = **9**

6) 9 − 6 = **3**

7) 8 − 7 = **1**

8) 9 − 2 = **7**

9) 7 − 3 = **4**

10) 6 − 1 = **5**

11) 10 − 8 = **2**

12) 10 − 9 = **1**

13) 7 − 6 = **1**

14) 10 − 7 = **3**

15) 10 − 9 = **1**

16) 8 − 1 = **7**

17) 9 − 4 = **5**

18) 6 − 6 = **0**

19) 10 − 7 = **3**

20) 7 − 6 = **1**

21) 9 − 7 = **2**

22) 5 − 1 = **4**

23) 10 − 4 = **6**

24) 3 − 2 = **1**

25) 7 − 6 = **1**

26) 6 − 2 = **4**

27) 4 − 2 = **2**

28) 10 − 9 = **1**

29) 9 − 1 = **8**

30) 2 − 1 = **1**

31) 5 − 2 = **3**

32) 8 − 4 = **4**

33) 9 − 3 = **6**

34) 9 − 3 = **6**

35) 1 − 0 = **1**

36) 7 − 6 = **1**

37) 10 − 0 = **10**

38) 8 − 5 = **3**

39) 9 − 2 = **7**

40) 3 − 2 = **1**

41) 10 − 3 = **7**

42) 9 − 7 = **2**

43) 7 − 1 = **6**

44) 3 − 1 = **2**

45) 10 − 3 = **7**

46) 9 − 1 = **8**

47) 9 − 5 = **4**

48) 10 − 6 = **4**

49) 10 − 1 = **9**

50) 7 − 6 = **1**

51) 8 − 4 = **4**

52) 9 − 7 = **2**

53) 7 − 3 = **4**

54) 3 − 3 = **0**

55) 8 − 5 = **3**

56) 8 − 3 = **5**

57) 10 − 9 = **1**

58) 10 − 3 = **7**

59) 3 − 1 = **2**

60) 3 − 2 = **1**

Date : _________

Time : ________

Name : __________

Score : _____/60

Subtraction Worksheet 0 - 10

1) 9 – 4 = _____

2) 5 – 3 = _____

3) 10 – 6 = _____

4) 8 – 1 = _____

5) 8 – 7 = _____

6) 9 – 1 = _____

7) 9 – 3 = _____

8) 10 – 4 = _____

9) 10 – 1 = _____

10) 7 – 6 = _____

11) 1 – 0 = _____

12) 2 – 0 = _____

13) 1 – 0 = _____

14) 7 – 5 = _____

15) 9 – 4 = _____

16) 4 – 1 = _____

17) 8 – 8 = _____

18) 7 – 4 = _____

19) 4 – 1 = _____

20) 10 – 6 = _____

21) 8 – 6 = _____

22) 4 – 3 = _____

23) 9 – 6 = _____

24) 4 – 1 = _____

25) 9 – 4 = _____

26) 10 – 7 = _____

27) 8 – 2 = _____

28) 7 – 3 = _____

29) 9 – 9 = _____

30) 8 – 5 = _____

31) 10 – 3 = _____

32) 7 – 1 = _____

33) 4 – 2 = _____

34) 3 – 1 = _____

35) 10 – 7 = _____

36) 6 – 1 = _____

37) 5 – 2 = _____

38) 6 – 1 = _____

39) 6 – 3 = _____

40) 3 – 3 = _____

41) 8 – 5 = _____

42) 8 – 2 = _____

43) 4 – 1 = _____

44) 8 – 5 = _____

45) 10 – 1 = _____

46) 3 – 2 = _____

47) 5 – 2 = _____

48) 8 – 4 = _____

49) 8 – 5 = _____

50) 7 – 3 = _____

51) 4 – 2 = _____

52) 9 – 7 = _____

53) 7 – 4 = _____

54) 7 – 4 = _____

55) 6 – 5 = _____

56) 9 – 4 = _____

57) 10 – 1 = _____

58) 10 – 2 = _____

59) 10 – 6 = _____

60) 7 – 7 = _____

Answer Key

Subtraction Worksheet 0 - 10

1) 9 – 4 = **5**   2) 5 – 3 = **2**   3) 10 – 6 = **4**

4) 8 – 1 = **7**   5) 8 – 7 = **1**   6) 9 – 1 = **8**

7) 9 – 3 = **6**   8) 10 – 4 = **6**   9) 10 – 1 = **9**

10) 7 – 6 = **1**   11) 1 – 0 = **1**   12) 2 – 0 = **2**

13) 1 – 0 = **1**   14) 7 – 5 = **2**   15) 9 – 4 = **5**

16) 4 – 1 = **3**   17) 8 – 8 = **0**   18) 7 – 4 = **3**

19) 4 – 1 = **3**   20) 10 – 6 = **4**   21) 8 – 6 = **2**

22) 4 – 3 = **1**   23) 9 – 6 = **3**   24) 4 – 1 = **3**

25) 9 – 4 = **5**   26) 10 – 7 = **3**   27) 8 – 2 = **6**

28) 7 – 3 = **4**   29) 9 – 9 = **0**   30) 8 – 5 = **3**

31) 10 – 3 = **7**   32) 7 – 1 = **6**   33) 4 – 2 = **2**

34) 3 – 1 = **2**   35) 10 – 7 = **3**   36) 6 – 1 = **5**

37) 5 – 2 = **3**   38) 6 – 1 = **5**   39) 6 – 3 = **3**

40) 3 – 3 = **0**   41) 8 – 5 = **3**   42) 8 – 2 = **6**

43) 4 – 1 = **3**   44) 8 – 5 = **3**   45) 10 – 1 = **9**

46) 3 – 2 = **1**   47) 5 – 2 = **3**   48) 8 – 4 = **4**

49) 8 – 5 = **3**   50) 7 – 3 = **4**   51) 4 – 2 = **2**

52) 9 – 7 = **2**   53) 7 – 4 = **3**   54) 7 – 4 = **3**

55) 6 – 5 = **1**   56) 9 – 4 = **5**   57) 10 – 1 = **9**

58) 10 – 2 = **8**   59) 10 – 6 = **4**   60) 7 – 7 = **0**

Date : _________   Name : _________
Time : _________   Score : _____/60

Subtraction Worksheet 0 - 10

1) 3 − 2 = _____   2) 6 − 4 = _____   3) 2 − 1 = _____

4) 6 − 1 = _____   5) 8 − 7 = _____   6) 8 − 1 = _____

7) 7 − 3 = _____   8) 9 − 3 = _____   9) 8 − 7 = _____

10) 6 − 2 = _____   11) 9 − 5 = _____   12) 7 − 2 = _____

13) 8 − 4 = _____   14) 9 − 4 = _____   15) 7 − 6 = _____

16) 6 − 2 = _____   17) 6 − 1 = _____   18) 8 − 3 = _____

19) 4 − 2 = _____   20) 5 − 3 = _____   21) 6 − 2 = _____

22) 7 − 3 = _____   23) 10 − 7 = _____   24) 10 − 5 = _____

25) 5 − 5 = _____   26) 4 − 3 = _____   27) 10 − 8 = _____

28) 10 − 9 = _____   29) 9 − 1 = _____   30) 5 − 2 = _____

31) 5 − 4 = _____   32) 6 − 5 = _____   33) 8 − 5 = _____

34) 8 − 4 = _____   35) 2 − 0 = _____   36) 9 − 7 = _____

37) 6 − 4 = _____   38) 7 − 6 = _____   39) 5 − 2 = _____

40) 9 − 6 = _____   41) 1 − 1 = _____   42) 5 − 4 = _____

43) 8 − 4 = _____   44) 4 − 4 = _____   45) 2 − 2 = _____

46) 3 − 3 = _____   47) 8 − 4 = _____   48) 10 − 9 = _____

49) 4 − 1 = _____   50) 8 − 5 = _____   51) 8 − 0 = _____

52) 10 − 5 = _____   53) 8 − 2 = _____   54) 5 − 2 = _____

55) 6 − 3 = _____   56) 4 − 2 = _____   57) 8 − 4 = _____

58) 7 − 5 = _____   59) 6 − 5 = _____   60) 6 − 2 = _____

Answer Key
Subtraction Worksheet 0 - 10

1) 3 – 2 = **1**

2) 6 – 4 = **2**

3) 2 – 1 = **1**

4) 6 – 1 = **5**

5) 8 – 7 = **1**

6) 8 – 1 = **7**

7) 7 – 3 = **4**

8) 9 – 3 = **6**

9) 8 – 7 = **1**

10) 6 – 2 = **4**

11) 9 – 5 = **4**

12) 7 – 2 = **5**

13) 8 – 4 = **4**

14) 9 – 4 = **5**

15) 7 – 6 = **1**

16) 6 – 2 = **4**

17) 6 – 1 = **5**

18) 8 – 3 = **5**

19) 4 – 2 = **2**

20) 5 – 3 = **2**

21) 6 – 2 = **4**

22) 7 – 3 = **4**

23) 10 – 7 = **3**

24) 10 – 5 = **5**

25) 5 – 5 = **0**

26) 4 – 3 = **1**

27) 10 – 8 = **2**

28) 10 – 9 = **1**

29) 9 – 1 = **8**

30) 5 – 2 = **3**

31) 5 – 4 = **1**

32) 6 – 5 = **1**

33) 8 – 5 = **3**

34) 8 – 4 = **4**

35) 2 – 0 = **2**

36) 9 – 7 = **2**

37) 6 – 4 = **2**

38) 7 – 6 = **1**

39) 5 – 2 = **3**

40) 9 – 6 = **3**

41) 1 – 1 = **0**

42) 5 – 4 = **1**

43) 8 – 4 = **4**

44) 4 – 4 = **0**

45) 2 – 2 = **0**

46) 3 – 3 = **0**

47) 8 – 4 = **4**

48) 10 – 9 = **1**

49) 4 – 1 = **3**

50) 8 – 5 = **3**

51) 8 – 0 = **8**

52) 10 – 5 = **5**

53) 8 – 2 = **6**

54) 5 – 2 = **3**

55) 6 – 3 = **3**

56) 4 – 2 = **2**

57) 8 – 4 = **4**

58) 7 – 5 = **2**

59) 6 – 5 = **1**

60) 6 – 2 = **4**

Subtraction Worksheet 0 - 20

1) 16 – 9 = _____   2) 20 – 8 = _____   3) 15 – 7 = _____

4) 5 – 3 = _____   5) 13 – 2 = _____   6) 11 – 1 = _____

7) 13 – 1 = _____   8) 18 – 10 = _____   9) 17 – 9 = _____

10) 14 – 12 = _____   11) 14 – 6 = _____   12) 19 – 8 = _____

13) 11 – 1 = _____   14) 6 – 1 = _____   15) 16 – 13 = _____

16) 11 – 8 = _____   17) 18 – 3 = _____   18) 11 – 6 = _____

19) 9 – 4 = _____   20) 11 – 8 = _____   21) 20 – 19 = _____

22) 13 – 4 = _____   23) 8 – 5 = _____   24) 13 – 2 = _____

25) 15 – 12 = _____   26) 14 – 12 = _____   27) 19 – 4 = _____

28) 17 – 1 = _____   29) 15 – 2 = _____   30) 14 – 12 = _____

31) 12 – 7 = _____   32) 15 – 5 = _____   33) 15 – 10 = _____

34) 15 – 15 = _____   35) 10 – 5 = _____   36) 14 – 6 = _____

37) 15 – 5 = _____   38) 20 – 13 = _____   39) 19 – 4 = _____

40) 12 – 3 = _____   41) 8 – 4 = _____   42) 16 – 5 = _____

43) 13 – 8 = _____   44) 20 – 12 = _____   45) 11 – 3 = _____

46) 6 – 1 = _____   47) 19 – 2 = _____   48) 9 – 9 = _____

49) 18 – 13 = _____   50) 16 – 1 = _____   51) 2 – 1 = _____

52) 18 – 7 = _____   53) 12 – 2 = _____   54) 15 – 1 = _____

55) 11 – 6 = _____   56) 18 – 15 = _____   57) 8 – 2 = _____

58) 15 – 13 = _____   59) 15 – 13 = _____   60) 17 – 11 = _____

Answer Key

Subtraction Worksheet 0 - 20

1) 16 – 9 = **7**

2) 20 – 8 = **12**

3) 15 – 7 = **8**

4) 5 – 3 = **2**

5) 13 – 2 = **11**

6) 11 – 1 = **10**

7) 13 – 1 = **12**

8) 18 – 10 = **8**

9) 17 – 9 = **8**

10) 14 – 12 = **2**

11) 14 – 6 = **8**

12) 19 – 8 = **11**

13) 11 – 1 = **10**

14) 6 – 1 = **5**

15) 16 – 13 = **3**

16) 11 – 8 = **3**

17) 18 – 3 = **15**

18) 11 – 6 = **5**

19) 9 – 4 = **5**

20) 11 – 8 = **3**

21) 20 – 19 = **1**

22) 13 – 4 = **9**

23) 8 – 5 = **3**

24) 13 – 2 = **11**

25) 15 – 12 = **3**

26) 14 – 12 = **2**

27) 19 – 4 = **15**

28) 17 – 1 = **16**

29) 15 – 2 = **13**

30) 14 – 12 = **2**

31) 12 – 7 = **5**

32) 15 – 5 = **10**

33) 15 – 10 = **5**

34) 15 – 15 = **0**

35) 10 – 5 = **5**

36) 14 – 6 = **8**

37) 15 – 5 = **10**

38) 20 – 13 = **7**

39) 19 – 4 = **15**

40) 12 – 3 = **9**

41) 8 – 4 = **4**

42) 16 – 5 = **11**

43) 13 – 8 = **5**

44) 20 – 12 = **8**

45) 11 – 3 = **8**

46) 6 – 1 = **5**

47) 19 – 2 = **17**

48) 9 – 9 = **0**

49) 18 – 13 = **5**

50) 16 – 1 = **15**

51) 2 – 1 = **1**

52) 18 – 7 = **11**

53) 12 – 2 = **10**

54) 15 – 1 = **14**

55) 11 – 6 = **5**

56) 18 – 15 = **3**

57) 8 – 2 = **6**

58) 15 – 13 = **2**

59) 15 – 13 = **2**

60) 17 – 11 = **6**

Date : _________

Time : _________

Name : _________

Score : _____/60

Subtraction Worksheet 0 - 20

1) 5 – 4 = _____

2) 13 – 6 = _____

3) 19 – 13 = _____

4) 18 – 14 = _____

5) 8 – 2 = _____

6) 10 – 3 = _____

7) 13 – 7 = _____

8) 15 – 4 = _____

9) 6 – 6 = _____

10) 17 – 11 = _____

11) 14 – 2 = _____

12) 19 – 17 = _____

13) 7 – 2 = _____

14) 19 – 8 = _____

15) 18 – 16 = _____

16) 11 – 11 = _____

17) 19 – 18 = _____

18) 13 – 4 = _____

19) 14 – 5 = _____

20) 15 – 1 = _____

21) 20 – 14 = _____

22) 12 – 9 = _____

23) 18 – 6 = _____

24) 16 – 4 = _____

25) 15 – 3 = _____

26) 19 – 10 = _____

27) 11 – 7 = _____

28) 19 – 3 = _____

29) 13 – 11 = _____

30) 13 – 8 = _____

31) 17 – 1 = _____

32) 18 – 1 = _____

33) 13 – 5 = _____

34) 14 – 7 = _____

35) 13 – 12 = _____

36) 17 – 7 = _____

37) 12 – 1 = _____

38) 16 – 4 = _____

39) 20 – 8 = _____

40) 14 – 12 = _____

41) 17 – 2 = _____

42) 12 – 4 = _____

43) 17 – 16 = _____

44) 14 – 12 = _____

45) 15 – 10 = _____

46) 17 – 17 = _____

47) 17 – 12 = _____

48) 16 – 14 = _____

49) 7 – 6 = _____

50) 16 – 12 = _____

51) 8 – 1 = _____

52) 13 – 10 = _____

53) 4 – 2 = _____

54) 13 – 3 = _____

55) 15 – 1 = _____

56) 11 – 8 = _____

57) 17 – 14 = _____

58) 13 – 9 = _____

59) 5 – 1 = _____

60) 16 – 0 = _____

Answer Key
Subtraction Worksheet 0 - 20

1) 5 – 4 = **1**	2) 13 – 6 = **7**	3) 19 – 13 = **6**
4) 18 – 14 = **4**	5) 8 – 2 = **6**	6) 10 – 3 = **7**
7) 13 – 7 = **6**	8) 15 – 4 = **11**	9) 6 – 6 = **0**
10) 17 – 11 = **6**	11) 14 – 2 = **12**	12) 19 – 17 = **2**
13) 7 – 2 = **5**	14) 19 – 8 = **11**	15) 18 – 16 = **2**
16) 11 – 11 = **0**	17) 19 – 18 = **1**	18) 13 – 4 = **9**
19) 14 – 5 = **9**	20) 15 – 1 = **14**	21) 20 – 14 = **6**
22) 12 – 9 = **3**	23) 18 – 6 = **12**	24) 16 – 4 = **12**
25) 15 – 3 = **12**	26) 19 – 10 = **9**	27) 11 – 7 = **4**
28) 19 – 3 = **16**	29) 13 – 11 = **2**	30) 13 – 8 = **5**
31) 17 – 1 = **16**	32) 18 – 1 = **17**	33) 13 – 5 = **8**
34) 14 – 7 = **7**	35) 13 – 12 = **1**	36) 17 – 7 = **10**
37) 12 – 1 = **11**	38) 16 – 4 = **12**	39) 20 – 8 = **12**
40) 14 – 12 = **2**	41) 17 – 2 = **15**	42) 12 – 4 = **8**
43) 17 – 16 = **1**	44) 14 – 12 = **2**	45) 15 – 10 = **5**
46) 17 – 17 = **0**	47) 17 – 12 = **5**	48) 16 – 14 = **2**
49) 7 – 6 = **1**	50) 16 – 12 = **4**	51) 8 – 1 = **7**
52) 13 – 10 = **3**	53) 4 – 2 = **2**	54) 13 – 3 = **10**
55) 15 – 1 = **14**	56) 11 – 8 = **3**	57) 17 – 14 = **3**
58) 13 – 9 = **4**	59) 5 – 1 = **4**	60) 16 – 0 = **16**

Subtraction Worksheet 0 - 20

1) 20 − 14 = _____   2) 16 − 2 = _____   3) 14 − 1 = _____

4) 5 − 2 = _____   5) 18 − 8 = _____   6) 11 − 4 = _____

7) 17 − 15 = _____   8) 13 − 0 = _____   9) 12 − 7 = _____

10) 8 − 3 = _____   11) 12 − 1 = _____   12) 19 − 3 = _____

13) 12 − 4 = _____   14) 20 − 13 = _____   15) 13 − 3 = _____

16) 6 − 2 = _____   17) 6 − 2 = _____   18) 10 − 3 = _____

19) 11 − 8 = _____   20) 18 − 4 = _____   21) 12 − 4 = _____

22) 6 − 4 = _____   23) 17 − 14 = _____   24) 10 − 3 = _____

25) 14 − 2 = _____   26) 14 − 5 = _____   27) 17 − 14 = _____

28) 15 − 2 = _____   29) 13 − 6 = _____   30) 16 − 7 = _____

31) 11 − 9 = _____   32) 20 − 14 = _____   33) 17 − 2 = _____

34) 19 − 16 = _____   35) 16 − 2 = _____   36) 20 − 10 = _____

37) 8 − 7 = _____   38) 9 − 7 = _____   39) 12 − 6 = _____

40) 9 − 0 = _____   41) 16 − 15 = _____   42) 20 − 20 = _____

43) 20 − 11 = _____   44) 12 − 12 = _____   45) 14 − 12 = _____

46) 20 − 5 = _____   47) 16 − 13 = _____   48) 17 − 15 = _____

49) 15 − 11 = _____   50) 15 − 15 = _____   51) 19 − 14 = _____

52) 14 − 8 = _____   53) 12 − 3 = _____   54) 6 − 4 = _____

55) 7 − 6 = _____   56) 18 − 12 = _____   57) 19 − 2 = _____

58) 20 − 12 = _____   59) 20 − 16 = _____   60) 11 − 9 = _____

Answer Key
Subtraction Worksheet 0 - 20

1) 20 − 14 = **6**

2) 16 − 2 = **14**

3) 14 − 1 = **13**

4) 5 − 2 = **3**

5) 18 − 8 = **10**

6) 11 − 4 = **7**

7) 17 − 15 = **2**

8) 13 − 0 = **13**

9) 12 − 7 = **5**

10) 8 − 3 = **5**

11) 12 − 1 = **11**

12) 19 − 3 = **16**

13) 12 − 4 = **8**

14) 20 − 13 = **7**

15) 13 − 3 = **10**

16) 6 − 2 = **4**

17) 6 − 2 = **4**

18) 10 − 3 = **7**

19) 11 − 8 = **3**

20) 18 − 4 = **14**

21) 12 − 4 = **8**

22) 6 − 4 = **2**

23) 17 − 14 = **3**

24) 10 − 3 = **7**

25) 14 − 2 = **12**

26) 14 − 5 = **9**

27) 17 − 14 = **3**

28) 15 − 2 = **13**

29) 13 − 6 = **7**

30) 16 − 7 = **9**

31) 11 − 9 = **2**

32) 20 − 14 = **6**

33) 17 − 2 = **15**

34) 19 − 16 = **3**

35) 16 − 2 = **14**

36) 20 − 10 = **10**

37) 8 − 7 = **1**

38) 9 − 7 = **2**

39) 12 − 6 = **6**

40) 9 − 0 = **9**

41) 16 − 15 = **1**

42) 20 − 20 = **0**

43) 20 − 11 = **9**

44) 12 − 12 = **0**

45) 14 − 12 = **2**

46) 20 − 5 = **15**

47) 16 − 13 = **3**

48) 17 − 15 = **2**

49) 15 − 11 = **4**

50) 15 − 15 = **0**

51) 19 − 14 = **5**

52) 14 − 8 = **6**

53) 12 − 3 = **9**

54) 6 − 4 = **2**

55) 7 − 6 = **1**

56) 18 − 12 = **6**

57) 19 − 2 = **17**

58) 20 − 12 = **8**

59) 20 − 16 = **4**

60) 11 − 9 = **2**

Date : _________

Time : _________

Name : __________

Score : _____/60

Subtraction Worksheet 0 - 20

1) 17 − 16 = _____   2) 14 − 1 = _____   3) 18 − 5 = _____

4) 4 − 2 = _____   5) 13 − 5 = _____   6) 13 − 4 = _____

7) 12 − 7 = _____   8) 11 − 11 = _____   9) 20 − 10 = _____

10) 16 − 14 = _____   11) 11 − 10 = _____   12) 13 − 7 = _____

13) 19 − 12 = _____   14) 18 − 16 = _____   15) 16 − 6 = _____

16) 15 − 14 = _____   17) 18 − 13 = _____   18) 14 − 6 = _____

19) 20 − 19 = _____   20) 20 − 16 = _____   21) 19 − 4 = _____

22) 18 − 9 = _____   23) 18 − 11 = _____   24) 17 − 10 = _____

25) 14 − 10 = _____   26) 13 − 9 = _____   27) 11 − 7 = _____

28) 14 − 4 = _____   29) 7 − 3 = _____   30) 17 − 14 = _____

31) 20 − 19 = _____   32) 18 − 9 = _____   33) 18 − 9 = _____

34) 14 − 7 = _____   35) 8 − 6 = _____   36) 14 − 14 = _____

37) 10 − 3 = _____   38) 9 − 3 = _____   39) 20 − 4 = _____

40) 16 − 14 = _____   41) 7 − 4 = _____   42) 16 − 5 = _____

43) 18 − 11 = _____   44) 19 − 14 = _____   45) 11 − 9 = _____

46) 12 − 1 = _____   47) 13 − 3 = _____   48) 9 − 6 = _____

49) 18 − 13 = _____   50) 20 − 11 = _____   51) 4 − 4 = _____

52) 17 − 0 = _____   53) 19 − 8 = _____   54) 7 − 0 = _____

55) 5 − 3 = _____   56) 13 − 5 = _____   57) 14 − 3 = _____

58) 15 − 9 = _____   59) 6 − 2 = _____   60) 20 − 16 = _____

Answer Key

Subtraction Worksheet 0 - 20

1) 17 – 16 = **1**

2) 14 – 1 = **13**

3) 18 – 5 = **13**

4) 4 – 2 = **2**

5) 13 – 5 = **8**

6) 13 – 4 = **9**

7) 12 – 7 = **5**

8) 11 – 11 = **0**

9) 20 – 10 = **10**

10) 16 – 14 = **2**

11) 11 – 10 = **1**

12) 13 – 7 = **6**

13) 19 – 12 = **7**

14) 18 – 16 = **2**

15) 16 – 6 = **10**

16) 15 – 14 = **1**

17) 18 – 13 = **5**

18) 14 – 6 = **8**

19) 20 – 19 = **1**

20) 20 – 16 = **4**

21) 19 – 4 = **15**

22) 18 – 9 = **9**

23) 18 – 11 = **7**

24) 17 – 10 = **7**

25) 14 – 10 = **4**

26) 13 – 9 = **4**

27) 11 – 7 = **4**

28) 14 – 4 = **10**

29) 7 – 3 = **4**

30) 17 – 14 = **3**

31) 20 – 19 = **1**

32) 18 – 9 = **9**

33) 18 – 9 = **9**

34) 14 – 7 = **7**

35) 8 – 6 = **2**

36) 14 – 14 = **0**

37) 10 – 3 = **7**

38) 9 – 3 = **6**

39) 20 – 4 = **16**

40) 16 – 14 = **2**

41) 7 – 4 = **3**

42) 16 – 5 = **11**

43) 18 – 11 = **7**

44) 19 – 14 = **5**

45) 11 – 9 = **2**

46) 12 – 1 = **11**

47) 13 – 3 = **10**

48) 9 – 6 = **3**

49) 18 – 13 = **5**

50) 20 – 11 = **9**

51) 4 – 4 = **0**

52) 17 – 0 = **17**

53) 19 – 8 = **11**

54) 7 – 0 = **7**

55) 5 – 3 = **2**

56) 13 – 5 = **8**

57) 14 – 3 = **11**

58) 15 – 9 = **6**

59) 6 – 2 = **4**

60) 20 – 16 = **4**

Date : _________

Time : _________

Name : __________

Score : _____/60

Subtraction Worksheet 0 - 20

1) 20 − 16 = _____   2) 4 − 2 = _____   3) 17 − 12 = _____

4) 18 − 13 = _____   5) 11 − 5 = _____   6) 12 − 0 = _____

7) 20 − 5 = _____   8) 16 − 6 = _____   9) 10 − 1 = _____

10) 19 − 7 = _____   11) 20 − 17 = _____   12) 15 − 12 = _____

13) 17 − 10 = _____   14) 16 − 6 = _____   15) 20 − 5 = _____

16) 10 − 8 = _____   17) 18 − 5 = _____   18) 13 − 1 = _____

19) 5 − 1 = _____   20) 19 − 16 = _____   21) 11 − 8 = _____

22) 9 − 5 = _____   23) 15 − 13 = _____   24) 12 − 8 = _____

25) 6 − 1 = _____   26) 3 − 3 = _____   27) 18 − 7 = _____

28) 11 − 2 = _____   29) 7 − 4 = _____   30) 9 − 7 = _____

31) 12 − 8 = _____   32) 13 − 5 = _____   33) 16 − 9 = _____

34) 17 − 15 = _____   35) 17 − 1 = _____   36) 20 − 19 = _____

37) 18 − 8 = _____   38) 13 − 2 = _____   39) 13 − 9 = _____

40) 19 − 9 = _____   41) 17 − 14 = _____   42) 12 − 2 = _____

43) 19 − 9 = _____   44) 17 − 10 = _____   45) 12 − 10 = _____

46) 16 − 6 = _____   47) 17 − 10 = _____   48) 10 − 1 = _____

49) 11 − 2 = _____   50) 17 − 15 = _____   51) 18 − 5 = _____

52) 14 − 3 = _____   53) 5 − 3 = _____   54) 18 − 11 = _____

55) 12 − 2 = _____   56) 18 − 5 = _____   57) 17 − 2 = _____

58) 11 − 2 = _____   59) 16 − 9 = _____   60) 17 − 5 = _____

Answer Key

Subtraction Worksheet 0 - 20

1) 20 − 16 = **4**

2) 4 − 2 = **2**

3) 17 − 12 = **5**

4) 18 − 13 = **5**

5) 11 − 5 = **6**

6) 12 − 0 = **12**

7) 20 − 5 = **15**

8) 16 − 6 = **10**

9) 10 − 1 = **9**

10) 19 − 7 = **12**

11) 20 − 17 = **3**

12) 15 − 12 = **3**

13) 17 − 10 = **7**

14) 16 − 6 = **10**

15) 20 − 5 = **15**

16) 10 − 8 = **2**

17) 18 − 5 = **13**

18) 13 − 1 = **12**

19) 5 − 1 = **4**

20) 19 − 16 = **3**

21) 11 − 8 = **3**

22) 9 − 5 = **4**

23) 15 − 13 = **2**

24) 12 − 8 = **4**

25) 6 − 1 = **5**

26) 3 − 3 = **0**

27) 18 − 7 = **11**

28) 11 − 2 = **9**

29) 7 − 4 = **3**

30) 9 − 7 = **2**

31) 12 − 8 = **4**

32) 13 − 5 = **8**

33) 16 − 9 = **7**

34) 17 − 15 = **2**

35) 17 − 1 = **16**

36) 20 − 19 = **1**

37) 18 − 8 = **10**

38) 13 − 2 = **11**

39) 13 − 9 = **4**

40) 19 − 9 = **10**

41) 17 − 14 = **3**

42) 12 − 2 = **10**

43) 19 − 9 = **10**

44) 17 − 10 = **7**

45) 12 − 10 = **2**

46) 16 − 6 = **10**

47) 17 − 10 = **7**

48) 10 − 1 = **9**

49) 11 − 2 = **9**

50) 17 − 15 = **2**

51) 18 − 5 = **13**

52) 14 − 3 = **11**

53) 5 − 3 = **2**

54) 18 − 11 = **7**

55) 12 − 2 = **10**

56) 18 − 5 = **13**

57) 17 − 2 = **15**

58) 11 − 2 = **9**

59) 16 − 9 = **7**

60) 17 − 5 = **12**

Date : _________   Name : __________

Time : ________   Score : _____/60

Subtraction Worksheet 0 - 20

1) 15 – 5 = _____   2) 16 – 10 = _____   3) 6 – 4 = _____

4) 6 – 3 = _____   5) 9 – 4 = _____   6) 12 – 1 = _____

7) 12 – 9 = _____   8) 13 – 9 = _____   9) 18 – 15 = _____

10) 19 – 19 = _____   11) 14 – 5 = _____   12) 10 – 4 = _____

13) 17 – 8 = _____   14) 4 – 1 = _____   15) 16 – 3 = _____

16) 12 – 2 = _____   17) 18 – 8 = _____   18) 19 – 10 = _____

19) 17 – 10 = _____   20) 20 – 9 = _____   21) 18 – 8 = _____

22) 20 – 1 = _____   23) 17 – 2 = _____   24) 13 – 10 = _____

25) 4 – 3 = _____   26) 20 – 16 = _____   27) 11 – 2 = _____

28) 11 – 6 = _____   29) 17 – 15 = _____   30) 19 – 11 = _____

31) 14 – 9 = _____   32) 16 – 9 = _____   33) 20 – 18 = _____

34) 11 – 5 = _____   35) 19 – 6 = _____   36) 12 – 6 = _____

37) 19 – 14 = _____   38) 14 – 7 = _____   39) 16 – 12 = _____

40) 10 – 9 = _____   41) 14 – 4 = _____   42) 12 – 3 = _____

43) 18 – 14 = _____   44) 9 – 5 = _____   45) 17 – 11 = _____

46) 14 – 1 = _____   47) 17 – 16 = _____   48) 18 – 10 = _____

49) 11 – 4 = _____   50) 17 – 0 = _____   51) 16 – 8 = _____

52) 17 – 3 = _____   53) 11 – 1 = _____   54) 11 – 6 = _____

55) 15 – 6 = _____   56) 12 – 3 = _____   57) 6 – 3 = _____

58) 9 – 3 = _____   59) 20 – 5 = _____   60) 2 – 1 = _____

Answer Key
Subtraction Worksheet 0 - 20

1) $15 - 5 = \mathbf{10}$	2) $16 - 10 = \mathbf{6}$	3) $6 - 4 = \mathbf{2}$
4) $6 - 3 = \mathbf{3}$	5) $9 - 4 = \mathbf{5}$	6) $12 - 1 = \mathbf{11}$
7) $12 - 9 = \mathbf{3}$	8) $13 - 9 = \mathbf{4}$	9) $18 - 15 = \mathbf{3}$
10) $19 - 19 = \mathbf{0}$	11) $14 - 5 = \mathbf{9}$	12) $10 - 4 = \mathbf{6}$
13) $17 - 8 = \mathbf{9}$	14) $4 - 1 = \mathbf{3}$	15) $16 - 3 = \mathbf{13}$
16) $12 - 2 = \mathbf{10}$	17) $18 - 8 = \mathbf{10}$	18) $19 - 10 = \mathbf{9}$
19) $17 - 10 = \mathbf{7}$	20) $20 - 9 = \mathbf{11}$	21) $18 - 8 = \mathbf{10}$
22) $20 - 1 = \mathbf{19}$	23) $17 - 2 = \mathbf{15}$	24) $13 - 10 = \mathbf{3}$
25) $4 - 3 = \mathbf{1}$	26) $20 - 16 = \mathbf{4}$	27) $11 - 2 = \mathbf{9}$
28) $11 - 6 = \mathbf{5}$	29) $17 - 15 = \mathbf{2}$	30) $19 - 11 = \mathbf{8}$
31) $14 - 9 = \mathbf{5}$	32) $16 - 9 = \mathbf{7}$	33) $20 - 18 = \mathbf{2}$
34) $11 - 5 = \mathbf{6}$	35) $19 - 6 = \mathbf{13}$	36) $12 - 6 = \mathbf{6}$
37) $19 - 14 = \mathbf{5}$	38) $14 - 7 = \mathbf{7}$	39) $16 - 12 = \mathbf{4}$
40) $10 - 9 = \mathbf{1}$	41) $14 - 4 = \mathbf{10}$	42) $12 - 3 = \mathbf{9}$
43) $18 - 14 = \mathbf{4}$	44) $9 - 5 = \mathbf{4}$	45) $17 - 11 = \mathbf{6}$
46) $14 - 1 = \mathbf{13}$	47) $17 - 16 = \mathbf{1}$	48) $18 - 10 = \mathbf{8}$
49) $11 - 4 = \mathbf{7}$	50) $17 - 0 = \mathbf{17}$	51) $16 - 8 = \mathbf{8}$
52) $17 - 3 = \mathbf{14}$	53) $11 - 1 = \mathbf{10}$	54) $11 - 6 = \mathbf{5}$
55) $15 - 6 = \mathbf{9}$	56) $12 - 3 = \mathbf{9}$	57) $6 - 3 = \mathbf{3}$
58) $9 - 3 = \mathbf{6}$	59) $20 - 5 = \mathbf{15}$	60) $2 - 1 = \mathbf{1}$

Date : ________

Time : ________

Name : _________

Score : _____/60

Subtraction Worksheet 0 - 20

1) 17 – 1 = _____

2) 15 – 7 = _____

3) 15 – 5 = _____

4) 10 – 5 = _____

5) 18 – 3 = _____

6) 20 – 19 = _____

7) 20 – 18 = _____

8) 20 – 13 = _____

9) 9 – 7 = _____

10) 6 – 5 = _____

11) 18 – 2 = _____

12) 14 – 8 = _____

13) 8 – 1 = _____

14) 9 – 4 = _____

15) 13 – 2 = _____

16) 18 – 10 = _____

17) 18 – 5 = _____

18) 17 – 14 = _____

19) 15 – 9 = _____

20) 18 – 15 = _____

21) 11 – 1 = _____

22) 10 – 1 = _____

23) 14 – 8 = _____

24) 12 – 2 = _____

25) 19 – 10 = _____

26) 6 – 5 = _____

27) 18 – 10 = _____

28) 17 – 14 = _____

29) 17 – 13 = _____

30) 17 – 14 = _____

31) 15 – 3 = _____

32) 19 – 3 = _____

33) 10 – 4 = _____

34) 3 – 2 = _____

35) 20 – 8 = _____

36) 14 – 7 = _____

37) 2 – 1 = _____

38) 10 – 6 = _____

39) 20 – 11 = _____

40) 18 – 15 = _____

41) 16 – 9 = _____

42) 15 – 12 = _____

43) 15 – 13 = _____

44) 14 – 5 = _____

45) 14 – 14 = _____

46) 9 – 9 = _____

47) 20 – 13 = _____

48) 13 – 3 = _____

49) 18 – 6 = _____

50) 19 – 2 = _____

51) 15 – 3 = _____

52) 14 – 10 = _____

53) 5 – 5 = _____

54) 9 – 2 = _____

55) 15 – 2 = _____

56) 10 – 6 = _____

57) 20 – 17 = _____

58) 16 – 8 = _____

59) 14 – 11 = _____

60) 20 – 7 = _____

Answer Key

Subtraction Worksheet 0 - 20

1) $17 - 1 = \mathbf{16}$

2) $15 - 7 = \mathbf{8}$

3) $15 - 5 = \mathbf{10}$

4) $10 - 5 = \mathbf{5}$

5) $18 - 3 = \mathbf{15}$

6) $20 - 19 = \mathbf{1}$

7) $20 - 18 = \mathbf{2}$

8) $20 - 13 = \mathbf{7}$

9) $9 - 7 = \mathbf{2}$

10) $6 - 5 = \mathbf{1}$

11) $18 - 2 = \mathbf{16}$

12) $14 - 8 = \mathbf{6}$

13) $8 - 1 = \mathbf{7}$

14) $9 - 4 = \mathbf{5}$

15) $13 - 2 = \mathbf{11}$

16) $18 - 10 = \mathbf{8}$

17) $18 - 5 = \mathbf{13}$

18) $17 - 14 = \mathbf{3}$

19) $15 - 9 = \mathbf{6}$

20) $18 - 15 = \mathbf{3}$

21) $11 - 1 = \mathbf{10}$

22) $10 - 1 = \mathbf{9}$

23) $14 - 8 = \mathbf{6}$

24) $12 - 2 = \mathbf{10}$

25) $19 - 10 = \mathbf{9}$

26) $6 - 5 = \mathbf{1}$

27) $18 - 10 = \mathbf{8}$

28) $17 - 14 = \mathbf{3}$

29) $17 - 13 = \mathbf{4}$

30) $17 - 14 = \mathbf{3}$

31) $15 - 3 = \mathbf{12}$

32) $19 - 3 = \mathbf{16}$

33) $10 - 4 = \mathbf{6}$

34) $3 - 2 = \mathbf{1}$

35) $20 - 8 = \mathbf{12}$

36) $14 - 7 = \mathbf{7}$

37) $2 - 1 = \mathbf{1}$

38) $10 - 6 = \mathbf{4}$

39) $20 - 11 = \mathbf{9}$

40) $18 - 15 = \mathbf{3}$

41) $16 - 9 = \mathbf{7}$

42) $15 - 12 = \mathbf{3}$

43) $15 - 13 = \mathbf{2}$

44) $14 - 5 = \mathbf{9}$

45) $14 - 14 = \mathbf{0}$

46) $9 - 9 = \mathbf{0}$

47) $20 - 13 = \mathbf{7}$

48) $13 - 3 = \mathbf{10}$

49) $18 - 6 = \mathbf{12}$

50) $19 - 2 = \mathbf{17}$

51) $15 - 3 = \mathbf{12}$

52) $14 - 10 = \mathbf{4}$

53) $5 - 5 = \mathbf{0}$

54) $9 - 2 = \mathbf{7}$

55) $15 - 2 = \mathbf{13}$

56) $10 - 6 = \mathbf{4}$

57) $20 - 17 = \mathbf{3}$

58) $16 - 8 = \mathbf{8}$

59) $14 - 11 = \mathbf{3}$

60) $20 - 7 = \mathbf{13}$

Date : _________ Name : _________

Time : _________ Score : _____/60

 Addition Missing Addends 0 - 20 

1) 0 + 20 = _____

2) 7 + _____ = 16

3) 14 + _____ = 18

4) _____ + 6 = 16

5) 9 + _____ = 16

6) 3 + _____ = 10

7) _____ + 7 = 19

8) _____ + 14 = 15

9) 8 + 11 = _____

10) 7 + _____ = 14

11) 3 + _____ = 7

12) 6 + 6 = _____

13) _____ + 10 = 17

14) 11 + _____ = 12

15) 10 + _____ = 11

16) _____ + 13 = 17

17) 9 + _____ = 19

18) 15 + 0 = _____

19) 18 + 1 = _____

20) 6 + 13 = _____

21) 1 + 2 = _____

22) _____ + 5 = 17

23) 2 + _____ = 10

24) 5 + _____ = 14

25) 17 + _____ = 18

26) 4 + _____ = 15

27) _____ + 12 = 19

28) 2 + _____ = 3

29) _____ + 5 = 12

30) 14 + _____ = 17

31) 16 + _____ = 18

32) 0 + _____ = 11

33) _____ + 4 = 18

34) _____ + 11 = 18

35) 7 + 12 = _____

36) 19 + _____ = 20

37) 11 + 3 = _____

38) 3 + _____ = 9

39) 2 + 18 = _____

40) 11 + _____ = 12

41) _____ + 14 = 19

42) _____ + 12 = 17

43) _____ + 1 = 12

44) 19 + _____ = 20

45) _____ + 1 = 17

46) _____ + 4 = 15

47) 11 + _____ = 16

48) _____ + 6 = 12

49) _____ + 5 = 15

50) _____ + 1 = 19

51) 1 + 13 = _____

52) 17 + 2 = _____

53) _____ + 8 = 18

54) 14 + 3 = _____

55) 7 + 9 = _____

56) _____ + 6 = 11

57) _____ + 13 = 19

58) _____ + 1 = 17

59) 10 + _____ = 13

60) _____ + 14 = 16

1) $0 + 20 = \mathbf{20}$	2) $7 + \mathbf{9} = 16$	3) $14 + \mathbf{4} = 18$
4) $\mathbf{10} + 6 = 16$	5) $9 + \mathbf{7} = 16$	6) $3 + \mathbf{7} = 10$
7) $\mathbf{12} + 7 = 19$	8) $\mathbf{1} + 14 = 15$	9) $8 + 11 = \mathbf{19}$
10) $7 + \mathbf{7} = 14$	11) $3 + \mathbf{4} = 7$	12) $6 + 6 = \mathbf{12}$
13) $\mathbf{7} + 10 = 17$	14) $11 + \mathbf{1} = 12$	15) $10 + \mathbf{1} = 11$
16) $\mathbf{4} + 13 = 17$	17) $9 + \mathbf{10} = 19$	18) $15 + 0 = \mathbf{15}$
19) $18 + 1 = \mathbf{19}$	20) $6 + 13 = \mathbf{19}$	21) $1 + 2 = \mathbf{3}$
22) $\mathbf{12} + 5 = 17$	23) $2 + \mathbf{8} = 10$	24) $5 + \mathbf{9} = 14$
25) $17 + \mathbf{1} = 18$	26) $4 + \mathbf{11} = 15$	27) $\mathbf{7} + 12 = 19$
28) $2 + \mathbf{1} = 3$	29) $\mathbf{7} + 5 = 12$	30) $14 + \mathbf{3} = 17$
31) $16 + \mathbf{2} = 18$	32) $0 + \mathbf{11} = 11$	33) $\mathbf{14} + 4 = 18$
34) $\mathbf{7} + 11 = 18$	35) $7 + 12 = \mathbf{19}$	36) $19 + \mathbf{1} = 20$
37) $11 + 3 = \mathbf{14}$	38) $3 + \mathbf{6} = 9$	39) $2 + 18 = \mathbf{20}$
40) $11 + \mathbf{1} = 12$	41) $\mathbf{5} + 14 = 19$	42) $\mathbf{5} + 12 = 17$
43) $\mathbf{11} + 1 = 12$	44) $19 + \mathbf{1} = 20$	45) $\mathbf{16} + 1 = 17$
46) $\mathbf{11} + 4 = 15$	47) $11 + \mathbf{5} = 16$	48) $\mathbf{6} + 6 = 12$
49) $\mathbf{10} + 5 = 15$	50) $\mathbf{18} + 1 = 19$	51) $1 + 13 = \mathbf{14}$
52) $17 + 2 = \mathbf{19}$	53) $\mathbf{10} + 8 = 18$	54) $14 + 3 = \mathbf{17}$
55) $7 + 9 = \mathbf{16}$	56) $\mathbf{5} + 6 = 11$	57) $\mathbf{6} + 13 = 19$
58) $\mathbf{16} + 1 = 17$	59) $10 + \mathbf{3} = 13$	60) $\mathbf{2} + 14 = 16$

Addition Missing Addends 0 - 20

1) $4 + \underline{\hspace{1cm}} = 14$

2) $5 + 14 = \underline{\hspace{1cm}}$

3) $0 + \underline{\hspace{1cm}} = 4$

4) $7 + 5 = \underline{\hspace{1cm}}$

5) $\underline{\hspace{1cm}} + 10 = 11$

6) $14 + \underline{\hspace{1cm}} = 20$

7) $20 + \underline{\hspace{1cm}} = 20$

8) $2 + 17 = \underline{\hspace{1cm}}$

9) $1 + \underline{\hspace{1cm}} = 16$

10) $7 + \underline{\hspace{1cm}} = 15$

11) $6 + \underline{\hspace{1cm}} = 20$

12) $2 + 18 = \underline{\hspace{1cm}}$

13) $19 + \underline{\hspace{1cm}} = 20$

14) $\underline{\hspace{1cm}} + 19 = 20$

15) $\underline{\hspace{1cm}} + 0 = 18$

16) $6 + 11 = \underline{\hspace{1cm}}$

17) $6 + \underline{\hspace{1cm}} = 19$

18) $1 + \underline{\hspace{1cm}} = 14$

19) $19 + \underline{\hspace{1cm}} = 20$

20) $15 + \underline{\hspace{1cm}} = 17$

21) $5 + 6 = \underline{\hspace{1cm}}$

22) $\underline{\hspace{1cm}} + 0 = 14$

23) $\underline{\hspace{1cm}} + 9 = 10$

24) $1 + \underline{\hspace{1cm}} = 15$

25) $15 + 2 = \underline{\hspace{1cm}}$

26) $7 + \underline{\hspace{1cm}} = 15$

27) $\underline{\hspace{1cm}} + 11 = 19$

28) $\underline{\hspace{1cm}} + 2 = 8$

29) $\underline{\hspace{1cm}} + 13 = 16$

30) $\underline{\hspace{1cm}} + 18 = 20$

31) $\underline{\hspace{1cm}} + 0 = 20$

32) $14 + \underline{\hspace{1cm}} = 18$

33) $4 + 7 = \underline{\hspace{1cm}}$

34) $15 + \underline{\hspace{1cm}} = 16$

35) $\underline{\hspace{1cm}} + 7 = 15$

36) $\underline{\hspace{1cm}} + 6 = 18$

37) $11 + \underline{\hspace{1cm}} = 15$

38) $\underline{\hspace{1cm}} + 2 = 19$

39) $\underline{\hspace{1cm}} + 13 = 17$

40) $1 + 11 = \underline{\hspace{1cm}}$

41) $\underline{\hspace{1cm}} + 5 = 14$

42) $6 + \underline{\hspace{1cm}} = 15$

43) $\underline{\hspace{1cm}} + 10 = 13$

44) $10 + 8 = \underline{\hspace{1cm}}$

45) $6 + \underline{\hspace{1cm}} = 8$

46) $2 + \underline{\hspace{1cm}} = 20$

47) $8 + \underline{\hspace{1cm}} = 8$

48) $2 + \underline{\hspace{1cm}} = 20$

49) $\underline{\hspace{1cm}} + 12 = 18$

50) $\underline{\hspace{1cm}} + 5 = 15$

51) $\underline{\hspace{1cm}} + 7 = 18$

52) $3 + \underline{\hspace{1cm}} = 19$

53) $1 + \underline{\hspace{1cm}} = 7$

54) $14 + \underline{\hspace{1cm}} = 15$

55) $\underline{\hspace{1cm}} + 9 = 11$

56) $\underline{\hspace{1cm}} + 3 = 18$

57) $5 + 15 = \underline{\hspace{1cm}}$

58) $\underline{\hspace{1cm}} + 20 = 20$

59) $8 + 12 = \underline{\hspace{1cm}}$

60) $1 + \underline{\hspace{1cm}} = 19$

Answer Key
Addition Missing Addends 0 - 20

1) 4 + **10** = 14

2) 5 + 14 = **19**

3) 0 + **4** = 4

4) 7 + 5 = **12**

5) **1** + 10 = 11

6) 14 + **6** = 20

7) 20 + **0** = 20

8) 2 + 17 = **19**

9) 1 + **15** = 16

10) 7 + **8** = 15

11) 6 + **14** = 20

12) 2 + 18 = **20**

13) 19 + **1** = 20

14) **1** + 19 = 20

15) **18** + 0 = 18

16) 6 + 11 = **17**

17) 6 + **13** = 19

18) 1 + **13** = 14

19) 19 + **1** = 20

20) 15 + **2** = 17

21) 5 + 6 = **11**

22) **14** + 0 = 14

23) **1** + 9 = 10

24) 1 + **14** = 15

25) 15 + 2 = **17**

26) 7 + **8** = 15

27) **8** + 11 = 19

28) **6** + 2 = 8

29) **3** + 13 = 16

30) **2** + 18 = 20

31) **20** + 0 = 20

32) 14 + **4** = 18

33) 4 + 7 = **11**

34) 15 + **1** = 16

35) **8** + 7 = 15

36) **12** + 6 = 18

37) 11 + **4** = 15

38) **17** + 2 = 19

39) **4** + 13 = 17

40) 1 + 11 = **12**

41) **9** + 5 = 14

42) 6 + **9** = 15

43) **3** + 10 = 13

44) 10 + 8 = **18**

45) 6 + **2** = 8

46) 2 + **18** = 20

47) 8 + **0** = 8

48) 2 + **18** = 20

49) **6** + 12 = 18

50) **10** + 5 = 15

51) **11** + 7 = 18

52) 3 + **16** = 19

53) 1 + **6** = 7

54) 14 + **1** = 15

55) **2** + 9 = 11

56) **15** + 3 = 18

57) 5 + 15 = **20**

58) **0** + 20 = 20

59) 8 + 12 = **20**

60) 1 + **18** = 19

 Addition Missing Addends 0 - 20 

1) $5 + \underline{\quad} = 7$

2) $\underline{\quad} + 18 = 19$

3) $11 + \underline{\quad} = 17$

4) $8 + \underline{\quad} = 9$

5) $11 + 8 = \underline{\quad}$

6) $\underline{\quad} + 4 = 16$

7) $\underline{\quad} + 1 = 17$

8) $17 + \underline{\quad} = 19$

9) $\underline{\quad} + 1 = 18$

10) $1 + \underline{\quad} = 15$

11) $3 + 7 = \underline{\quad}$

12) $4 + \underline{\quad} = 9$

13) $\underline{\quad} + 1 = 9$

14) $5 + \underline{\quad} = 20$

15) $\underline{\quad} + 13 = 16$

16) $8 + \underline{\quad} = 17$

17) $\underline{\quad} + 14 = 19$

18) $15 + \underline{\quad} = 18$

19) $\underline{\quad} + 19 = 20$

20) $4 + \underline{\quad} = 18$

21) $\underline{\quad} + 5 = 20$

22) $\underline{\quad} + 13 = 17$

23) $\underline{\quad} + 6 = 7$

24) $17 + 1 = \underline{\quad}$

25) $19 + 1 = \underline{\quad}$

26) $\underline{\quad} + 9 = 19$

27) $\underline{\quad} + 9 = 14$

28) $12 + 7 = \underline{\quad}$

29) $17 + 3 = \underline{\quad}$

30) $11 + 7 = \underline{\quad}$

31) $11 + 9 = \underline{\quad}$

32) $19 + 0 = \underline{\quad}$

33) $\underline{\quad} + 14 = 17$

34) $\underline{\quad} + 12 = 13$

35) $\underline{\quad} + 3 = 15$

36) $\underline{\quad} + 7 = 9$

37) $10 + 2 = \underline{\quad}$

38) $10 + 10 = \underline{\quad}$

39) $5 + 14 = \underline{\quad}$

40) $19 + \underline{\quad} = 19$

41) $4 + \underline{\quad} = 7$

42) $\underline{\quad} + 16 = 20$

43) $1 + 14 = \underline{\quad}$

44) $\underline{\quad} + 16 = 17$

45) $3 + 5 = \underline{\quad}$

46) $15 + \underline{\quad} = 19$

47) $\underline{\quad} + 2 = 3$

48) $\underline{\quad} + 5 = 9$

49) $\underline{\quad} + 1 = 20$

50) $\underline{\quad} + 11 = 17$

51) $3 + \underline{\quad} = 18$

52) $\underline{\quad} + 2 = 4$

53) $4 + 12 = \underline{\quad}$

54) $19 + 1 = \underline{\quad}$

55) $12 + \underline{\quad} = 19$

56) $2 + \underline{\quad} = 18$

57) $17 + \underline{\quad} = 19$

58) $10 + \underline{\quad} = 16$

59) $\underline{\quad} + 1 = 20$

60) $\underline{\quad} + 4 = 19$

Answer Key
Addition Missing Addends 0 - 20

1) $5 + \mathbf{2} = 7$

2) $\mathbf{1} + 18 = 19$

3) $11 + \mathbf{6} = 17$

4) $8 + \mathbf{1} = 9$

5) $11 + 8 = \mathbf{19}$

6) $\mathbf{12} + 4 = 16$

7) $\mathbf{16} + 1 = 17$

8) $17 + \mathbf{2} = 19$

9) $\mathbf{17} + 1 = 18$

10) $1 + \mathbf{14} = 15$

11) $3 + 7 = \mathbf{10}$

12) $4 + \mathbf{5} = 9$

13) $\mathbf{8} + 1 = 9$

14) $5 + \mathbf{15} = 20$

15) $\mathbf{3} + 13 = 16$

16) $8 + \mathbf{9} = 17$

17) $\mathbf{5} + 14 = 19$

18) $15 + \mathbf{3} = 18$

19) $\mathbf{1} + 19 = 20$

20) $4 + \mathbf{14} = 18$

21) $\mathbf{15} + 5 = 20$

22) $\mathbf{4} + 13 = 17$

23) $\mathbf{1} + 6 = 7$

24) $17 + 1 = \mathbf{18}$

25) $19 + 1 = \mathbf{20}$

26) $\mathbf{10} + 9 = 19$

27) $\mathbf{5} + 9 = 14$

28) $12 + 7 = \mathbf{19}$

29) $17 + 3 = \mathbf{20}$

30) $11 + 7 = \mathbf{18}$

31) $11 + 9 = \mathbf{20}$

32) $19 + 0 = \mathbf{19}$

33) $\mathbf{3} + 14 = 17$

34) $\mathbf{1} + 12 = 13$

35) $\mathbf{12} + 3 = 15$

36) $\mathbf{2} + 7 = 9$

37) $10 + 2 = \mathbf{12}$

38) $10 + 10 = \mathbf{20}$

39) $5 + 14 = \mathbf{19}$

40) $19 + \mathbf{0} = 19$

41) $4 + \mathbf{3} = 7$

42) $\mathbf{4} + 16 = 20$

43) $1 + 14 = \mathbf{15}$

44) $\mathbf{1} + 16 = 17$

45) $3 + 5 = \mathbf{8}$

46) $15 + \mathbf{4} = 19$

47) $\mathbf{1} + 2 = 3$

48) $\mathbf{4} + 5 = 9$

49) $\mathbf{19} + 1 = 20$

50) $\mathbf{6} + 11 = 17$

51) $3 + \mathbf{15} = 18$

52) $\mathbf{2} + 2 = 4$

53) $4 + 12 = \mathbf{16}$

54) $19 + 1 = \mathbf{20}$

55) $12 + \mathbf{7} = 19$

56) $2 + \mathbf{16} = 18$

57) $17 + \mathbf{2} = 19$

58) $10 + \mathbf{6} = 16$

59) $\mathbf{19} + 1 = 20$

60) $\mathbf{15} + 4 = 19$

Date : _________

Time : _________

Name : _________

Score : _____/60

Addition Missing Addends 0 - 20

1) 18 + _____ = 19

2) _____ + 17 = 18

3) _____ + 18 = 20

4) _____ + 7 = 18

5) 12 + 6 = _____

6) 4 + _____ = 16

7) _____ + 10 = 18

8) _____ + 13 = 17

9) _____ + 3 = 6

10) _____ + 8 = 13

11) _____ + 2 = 20

12) _____ + 2 = 20

13) _____ + 14 = 15

14) 1 + _____ = 11

15) _____ + 15 = 17

16) 5 + _____ = 8

17) 9 + _____ = 15

18) _____ + 8 = 13

19) _____ + 0 = 20

20) _____ + 2 = 18

21) _____ + 5 = 9

22) _____ + 1 = 7

23) 14 + 2 = _____

24) 19 + _____ = 20

25) 10 + _____ = 19

26) 13 + 6 = _____

27) 8 + 6 = _____

28) _____ + 6 = 10

29) _____ + 15 = 15

30) 2 + _____ = 17

31) 1 + _____ = 20

32) _____ + 3 = 16

33) 0 + 18 = _____

34) _____ + 18 = 20

35) _____ + 9 = 20

36) 2 + _____ = 10

37) 8 + _____ = 11

38) 6 + 2 = _____

39) 5 + 14 = _____

40) 13 + _____ = 18

41) 4 + _____ = 9

42) 3 + _____ = 7

43) 15 + _____ = 18

44) 6 + _____ = 17

45) _____ + 2 = 10

46) _____ + 10 = 19

47) _____ + 2 = 20

48) _____ + 3 = 12

49) 2 + 18 = _____

50) 9 + _____ = 19

51) _____ + 0 = 9

52) 19 + _____ = 20

53) _____ + 8 = 17

54) 5 + _____ = 13

55) 0 + _____ = 20

56) 11 + 0 = _____

57) 6 + 15 = _____

58) 19 + _____ = 19

59) 11 + 1 = _____

60) _____ + 5 = 14

Answer Key
Addition Missing Addends 0 - 20

1) 18 + **1** = 19

2) **1** + 17 = 18

3) **2** + 18 = 20

4) **11** + 7 = 18

5) 12 + 6 = **18**

6) 4 + **12** = 16

7) **8** + 10 = 18

8) **4** + 13 = 17

9) **3** + 3 = 6

10) **5** + 8 = 13

11) **18** + 2 = 20

12) **18** + 2 = 20

13) **1** + 14 = 15

14) 1 + **10** = 11

15) **2** + 15 = 17

16) 5 + **3** = 8

17) 9 + **6** = 15

18) **5** + 8 = 13

19) **20** + 0 = 20

20) **16** + 2 = 18

21) **4** + 5 = 9

22) **6** + 1 = 7

23) 14 + 2 = **16**

24) 19 + **1** = 20

25) 10 + **9** = 19

26) 13 + 6 = **19**

27) 8 + 6 = **14**

28) **4** + 6 = 10

29) **0** + 15 = 15

30) 2 + **15** = 17

31) 1 + **19** = 20

32) **13** + 3 = 16

33) 0 + 18 = **18**

34) **2** + 18 = 20

35) **11** + 9 = 20

36) 2 + **8** = 10

37) 8 + **3** = 11

38) 6 + 2 = **8**

39) 5 + 14 = **19**

40) 13 + **5** = 18

41) 4 + **5** = 9

42) 3 + **4** = 7

43) 15 + **3** = 18

44) 6 + **11** = 17

45) **8** + 2 = 10

46) **9** + 10 = 19

47) **18** + 2 = 20

48) **9** + 3 = 12

49) 2 + 18 = **20**

50) 9 + **10** = 19

51) **9** + 0 = 9

52) 19 + **1** = 20

53) **9** + 8 = 17

54) 5 + **8** = 13

55) 0 + **20** = 20

56) 11 + 0 = **11**

57) 6 + 15 = **21**

58) 19 + **0** = 19

59) 11 + 1 = **12**

60) **9** + 5 = 14

 Addition Missing Addends 0 - 20 

1) _____ + 1 = 14

2) _____ + 1 = 17

3) 3 + _____ = 19

4) 0 + 19 = _____

5) 9 + 5 = _____

6) 11 + _____ = 16

7) 14 + 5 = _____

8) _____ + 1 = 20

9) 2 + _____ = 9

10) 6 + 13 = _____

11) _____ + 3 = 18

12) 4 + 7 = _____

13) _____ + 2 = 20

14) _____ + 17 = 20

15) _____ + 18 = 19

16) _____ + 9 = 13

17) 8 + _____ = 10

18) 16 + 4 = _____

19) 4 + _____ = 11

20) 12 + _____ = 17

21) 10 + 4 = _____

22) 9 + _____ = 15

23) _____ + 15 = 19

24) 2 + 7 = _____

25) 0 + _____ = 18

26) _____ + 4 = 14

27) 2 + 10 = _____

28) 12 + _____ = 18

29) 9 + 6 = _____

30) _____ + 2 = 11

31) 1 + _____ = 20

32) _____ + 5 = 18

33) _____ + 3 = 17

34) _____ + 20 = 21

35) 17 + _____ = 20

36) 13 + 3 = _____

37) 6 + _____ = 9

38) 1 + _____ = 9

39) 4 + 9 = _____

40) 14 + 6 = _____

41) _____ + 1 = 15

42) 16 + 2 = _____

43) 0 + _____ = 9

44) 0 + _____ = 20

45) _____ + 6 = 9

46) _____ + 6 = 9

47) 5 + _____ = 10

48) 10 + _____ = 19

49) _____ + 8 = 8

50) 6 + 0 = _____

51) 1 + _____ = 7

52) _____ + 10 = 19

53) _____ + 3 = 11

54) 14 + _____ = 20

55) 3 + 7 = _____

56) 15 + 2 = _____

57) 20 + 0 = _____

58) 7 + _____ = 15

59) 16 + _____ = 18

60) 5 + 7 = _____

Answer Key
Addition Missing Addends 0 - 20

1) **13** + 1 = 14

2) **16** + 1 = 17

3) 3 + **16** = 19

4) 0 + 19 = **19**

5) 9 + 5 = **14**

6) 11 + **5** = 16

7) 14 + 5 = **19**

8) **19** + 1 = 20

9) 2 + **7** = 9

10) 6 + 13 = **19**

11) **15** + 3 = 18

12) 4 + 7 = **11**

13) **18** + 2 = 20

14) **3** + 17 = 20

15) **1** + 18 = 19

16) **4** + 9 = 13

17) 8 + **2** = 10

18) 16 + 4 = **20**

19) 4 + **7** = 11

20) 12 + **5** = 17

21) 10 + 4 = **14**

22) 9 + **6** = 15

23) **4** + 15 = 19

24) 2 + 7 = **9**

25) 0 + **18** = 18

26) **10** + 4 = 14

27) 2 + 10 = **12**

28) 12 + **6** = 18

29) 9 + 6 = **15**

30) **9** + 2 = 11

31) 1 + **19** = 20

32) **13** + 5 = 18

33) **14** + 3 = 17

34) **1** + 20 = 21

35) 17 + **3** = 20

36) 13 + 3 = **16**

37) 6 + **3** = 9

38) 1 + **8** = 9

39) 4 + 9 = **13**

40) 14 + 6 = **20**

41) **14** + 1 = 15

42) 16 + 2 = **18**

43) 0 + **9** = 9

44) 0 + **20** = 20

45) **3** + 6 = 9

46) **3** + 6 = 9

47) 5 + **5** = 10

48) 10 + **9** = 19

49) **0** + 8 = 8

50) 6 + 0 = **6**

51) 1 + **6** = 7

52) **9** + 10 = 19

53) **8** + 3 = 11

54) 14 + **6** = 20

55) 3 + 7 = **10**

56) 15 + 2 = **17**

57) 20 + 0 = **20**

58) 7 + **8** = 15

59) 16 + **2** = 18

60) 5 + 7 = **12**

Addition Missing Addends 0 - 20

1) _____ + 6 = 7

2) 5 + 12 = _____

3) _____ + 12 = 15

4) 2 + _____ = 17

5) _____ + 0 = 18

6) 7 + _____ = 8

7) 2 + _____ = 15

8) 12 + _____ = 16

9) 20 + _____ = 20

10) _____ + 15 = 19

11) _____ + 1 = 18

12) _____ + 14 = 15

13) _____ + 18 = 20

14) 1 + _____ = 20

15) 1 + _____ = 20

16) 9 + _____ = 18

17) 13 + 2 = _____

18) _____ + 2 = 8

19) 20 + 0 = _____

20) 3 + 17 = _____

21) 3 + _____ = 10

22) _____ + 6 = 18

23) 10 + 2 = _____

24) _____ + 8 = 14

25) 1 + 6 = _____

26) 1 + _____ = 20

27) 17 + _____ = 18

28) 19 + 1 = _____

29) 7 + _____ = 18

30) _____ + 4 = 4

31) 17 + _____ = 20

32) 1 + _____ = 10

33) 19 + _____ = 19

34) _____ + 19 = 20

35) 9 + 8 = _____

36) 1 + 11 = _____

37) _____ + 0 = 19

38) _____ + 2 = 15

39) 12 + _____ = 16

40) 6 + _____ = 20

41) 17 + 1 = _____

42) 14 + _____ = 15

43) 1 + 19 = _____

44) _____ + 4 = 20

45) 2 + _____ = 3

46) 17 + _____ = 19

47) 6 + _____ = 10

48) _____ + 5 = 14

49) 16 + 4 = _____

50) _____ + 7 = 18

51) 5 + _____ = 17

52) 1 + _____ = 15

53) 4 + _____ = 19

54) _____ + 2 = 16

55) 2 + _____ = 18

56) 8 + _____ = 13

57) 13 + 6 = _____

58) _____ + 9 = 16

59) 1 + _____ = 12

60) 11 + _____ = 18

Answer Key
Addition Missing Addends 0 - 20

1) **1** + 6 = 7

2) 5 + 12 = **17**

3) **3** + 12 = 15

4) 2 + **15** = 17

5) **18** + 0 = 18

6) 7 + **1** = 8

7) 2 + **13** = 15

8) 12 + **4** = 16

9) 20 + **0** = 20

10) **4** + 15 = 19

11) **17** + 1 = 18

12) **1** + 14 = 15

13) **2** + 18 = 20

14) 1 + **19** = 20

15) 1 + **19** = 20

16) 9 + **9** = 18

17) 13 + 2 = **15**

18) **6** + 2 = 8

19) 20 + 0 = **20**

20) 3 + 17 = **20**

21) 3 + **7** = 10

22) **12** + 6 = 18

23) 10 + 2 = **12**

24) **6** + 8 = 14

25) 1 + 6 = **7**

26) 1 + **19** = 20

27) 17 + **1** = 18

28) 19 + 1 = **20**

29) 7 + **11** = 18

30) **0** + 4 = 4

31) 17 + **3** = 20

32) 1 + **9** = 10

33) 19 + **0** = 19

34) **1** + 19 = 20

35) 9 + 8 = **17**

36) 1 + 11 = **12**

37) **19** + 0 = 19

38) **13** + 2 = 15

39) 12 + **4** = 16

40) 6 + **14** = 20

41) 17 + 1 = **18**

42) 14 + **1** = 15

43) 1 + 19 = **20**

44) **16** + 4 = 20

45) 2 + **1** = 3

46) 17 + **2** = 19

47) 6 + **4** = 10

48) **9** + 5 = 14

49) 16 + 4 = **20**

50) **11** + 7 = 18

51) 5 + **12** = 17

52) 1 + **14** = 15

53) 4 + **15** = 19

54) **14** + 2 = 16

55) 2 + **16** = 18

56) 8 + **5** = 13

57) 13 + 6 = **19**

58) **7** + 9 = 16

59) 1 + **11** = 12

60) 11 + **7** = 18

Addition Missing Addends 0 - 20

1) $6 + \underline{\quad} = 16$

2) $\underline{\quad} + 15 = 16$

3) $10 + 5 = \underline{\quad}$

4) $3 + \underline{\quad} = 20$

5) $0 + \underline{\quad} = 19$

6) $17 + \underline{\quad} = 18$

7) $\underline{\quad} + 12 = 19$

8) $\underline{\quad} + 17 = 18$

9) $\underline{\quad} + 13 = 14$

10) $4 + \underline{\quad} = 5$

11) $1 + 19 = \underline{\quad}$

12) $3 + \underline{\quad} = 7$

13) $\underline{\quad} + 14 = 20$

14) $12 + \underline{\quad} = 19$

15) $\underline{\quad} + 18 = 20$

16) $\underline{\quad} + 20 = 20$

17) $6 + \underline{\quad} = 19$

18) $16 + \underline{\quad} = 19$

19) $\underline{\quad} + 10 = 17$

20) $\underline{\quad} + 20 = 20$

21) $17 + \underline{\quad} = 19$

22) $10 + \underline{\quad} = 19$

23) $\underline{\quad} + 10 = 21$

24) $\underline{\quad} + 5 = 19$

25) $7 + 6 = \underline{\quad}$

26) $3 + \underline{\quad} = 12$

27) $\underline{\quad} + 1 = 13$

28) $5 + \underline{\quad} = 20$

29) $5 + \underline{\quad} = 6$

30) $4 + \underline{\quad} = 7$

31) $2 + \underline{\quad} = 4$

32) $6 + 9 = \underline{\quad}$

33) $13 + \underline{\quad} = 14$

34) $\underline{\quad} + 0 = 19$

35) $14 + \underline{\quad} = 19$

36) $\underline{\quad} + 4 = 18$

37) $11 + \underline{\quad} = 20$

38) $15 + \underline{\quad} = 18$

39) $16 + 2 = \underline{\quad}$

40) $\underline{\quad} + 1 = 18$

41) $15 + 3 = \underline{\quad}$

42) $\underline{\quad} + 18 = 18$

43) $\underline{\quad} + 20 = 20$

44) $\underline{\quad} + 12 = 19$

45) $1 + \underline{\quad} = 11$

46) $5 + \underline{\quad} = 18$

47) $\underline{\quad} + 4 = 13$

48) $4 + \underline{\quad} = 18$

49) $15 + \underline{\quad} = 17$

50) $\underline{\quad} + 18 = 18$

51) $\underline{\quad} + 6 = 15$

52) $10 + 5 = \underline{\quad}$

53) $14 + 1 = \underline{\quad}$

54) $7 + \underline{\quad} = 19$

55) $\underline{\quad} + 13 = 19$

56) $\underline{\quad} + 19 = 20$

57) $\underline{\quad} + 19 = 20$

58) $8 + \underline{\quad} = 19$

59) $\underline{\quad} + 13 = 16$

60) $7 + 8 = \underline{\quad}$

Answer Key
Addition Missing Addends 0 - 20

1) 6 + **10** = 16

2) **1** + 15 = 16

3) 10 + 5 = **15**

4) 3 + **17** = 20

5) 0 + **19** = 19

6) 17 + **1** = 18

7) **7** + 12 = 19

8) **1** + 17 = 18

9) **1** + 13 = 14

10) 4 + **1** = 5

11) 1 + 19 = **20**

12) 3 + **4** = 7

13) **6** + 14 = 20

14) 12 + **7** = 19

15) **2** + 18 = 20

16) **0** + 20 = 20

17) 6 + **13** = 19

18) 16 + **3** = 19

19) **7** + 10 = 17

20) **0** + 20 = 20

21) 17 + **2** = 19

22) 10 + **9** = 19

23) **11** + 10 = 21

24) **14** + 5 = 19

25) 7 + 6 = **13**

26) 3 + **9** = 12

27) **12** + 1 = 13

28) 5 + **15** = 20

29) 5 + **1** = 6

30) 4 + **3** = 7

31) 2 + **2** = 4

32) 6 + 9 = **15**

33) 13 + **1** = 14

34) **19** + 0 = 19

35) 14 + **5** = 19

36) **14** + 4 = 18

37) 11 + **9** = 20

38) 15 + **3** = 18

39) 16 + 2 = **18**

40) **17** + 1 = 18

41) 15 + 3 = **18**

42) **0** + 18 = 18

43) **0** + 20 = 20

44) **7** + 12 = 19

45) 1 + **10** = 11

46) 5 + **13** = 18

47) **9** + 4 = 13

48) 4 + **14** = 18

49) 15 + **2** = 17

50) **0** + 18 = 18

51) **9** + 6 = 15

52) 10 + 5 = **15**

53) 14 + 1 = **15**

54) 7 + **12** = 19

55) **6** + 13 = 19

56) **1** + 19 = 20

57) **1** + 19 = 20

58) 8 + **11** = 19

59) **3** + 13 = 16

60) 7 + 8 = **15**

Subtraction Missing Addends 0 - 20

1) _____ − 9 = 7

2) 16 − 10 = _____

3) 15 − 9 = _____

4) 18 − _____ = 17

5) _____ − 4 = 9

6) _____ − 7 = 8

7) 19 − _____ = 17

8) _____ − 8 = 8

9) 14 − 12 = _____

10) 6 − _____ = 5

11) _____ − 9 = 2

12) 17 − 3 = _____

13) _____ − 17 = 1

14) _____ − 5 = 8

15) _____ − 13 = 4

16) _____ − 2 = 1

17) 18 − _____ = 4

18) _____ − 4 = 12

19) 14 − _____ = 9

20) 18 − 7 = _____

21) 8 − _____ = 6

22) 6 − 0 = _____

23) 18 − _____ = 10

24) 3 − _____ = 0

25) _____ − 7 = 10

26) 8 − 2 = _____

27) _____ − 3 = 2

28) _____ − 6 = 11

29) _____ − 14 = 1

30) _____ − 4 = 10

31) _____ − 9 = 7

32) _____ − 1 = 4

33) 18 − 12 = _____

34) 14 − 8 = _____

35) _____ − 11 = 4

36) 18 − 17 = _____

37) 20 − _____ = 4

38) 13 − _____ = 1

39) 15 − _____ = 6

40) _____ − 2 = 17

41) 8 − _____ = 3

42) _____ − 3 = 4

43) _____ − 1 = 15

44) 10 − _____ = 5

45) _____ − 12 = 8

46) _____ − 14 = 1

47) 6 − 4 = _____

48) 14 − _____ = 9

49) 15 − _____ = 12

50) 16 − _____ = 11

51) 12 − 5 = _____

52) 9 − _____ = 3

53) 8 − 3 = _____

54) 17 − 2 = _____

55) _____ − 3 = 16

56) 17 − _____ = 16

57) _____ − 2 = 7

58) 12 − _____ = 10

59) 11 − _____ = 9

60) 12 − 10 = _____

Answer Key
Subtraction Missing Addends 0 - 20

1) **16** − 9 = 7
2) 16 − 10 = **6**
3) 15 − 9 = **6**
4) 18 − **1** = 17
5) **13** − 4 = 9
6) **15** − 7 = 8
7) 19 − **2** = 17
8) **16** − 8 = 8
9) 14 − 12 = **2**
10) 6 − **1** = 5
11) **11** − 9 = 2
12) 17 − 3 = **14**
13) **18** − 17 = 1
14) **13** − 5 = 8
15) **17** − 13 = 4
16) **3** − 2 = 1
17) 18 − **14** = 4
18) **16** − 4 = 12
19) 14 − **5** = 9
20) 18 − 7 = **11**
21) 8 − **2** = 6
22) 6 − 0 = **6**
23) 18 − **8** = 10
24) 3 − **3** = 0
25) **17** − 7 = 10
26) 8 − 2 = **6**
27) **5** − 3 = 2
28) **17** − 6 = 11
29) **15** − 14 = 1
30) **14** − 4 = 10
31) **16** − 9 = 7
32) **5** − 1 = 4
33) 18 − 12 = **6**
34) 14 − 8 = **6**
35) **15** − 11 = 4
36) 18 − 17 = **1**
37) 20 − **16** = 4
38) 13 − **12** = 1
39) 15 − **9** = 6
40) **19** − 2 = 17
41) 8 − **5** = 3
42) **7** − 3 = 4
43) **16** − 1 = 15
44) 10 − **5** = 5
45) **20** − 12 = 8
46) **15** − 14 = 1
47) 6 − 4 = **2**
48) 14 − **5** = 9
49) 15 − **3** = 12
50) 16 − **5** = 11
51) 12 − 5 = **7**
52) 9 − **6** = 3
53) 8 − 3 = **5**
54) 17 − 2 = **15**
55) **19** − 3 = 16
56) 17 − **1** = 16
57) **9** − 2 = 7
58) 12 − **2** = 10
59) 11 − **2** = 9
60) 12 − 10 = **2**

 Subtraction Missing Addends 0 - 20 

1) $16 - \underline{\quad} = 13$

2) $15 - \underline{\quad} = 2$

3) $14 - \underline{\quad} = 5$

4) $19 - \underline{\quad} = 10$

5) $6 - \underline{\quad} = 1$

6) $18 - \underline{\quad} = 15$

7) $18 - \underline{\quad} = 2$

8) $\underline{\quad} - 4 = 11$

9) $\underline{\quad} - 15 = 4$

10) $12 - 7 = \underline{\quad}$

11) $19 - \underline{\quad} = 10$

12) $12 - \underline{\quad} = 9$

13) $9 - 5 = \underline{\quad}$

14) $14 - 10 = \underline{\quad}$

15) $12 - 4 = \underline{\quad}$

16) $\underline{\quad} - 14 = 5$

17) $10 - 7 = \underline{\quad}$

18) $\underline{\quad} - 1 = 6$

19) $13 - 12 = \underline{\quad}$

20) $\underline{\quad} - 0 = 8$

21) $11 - 6 = \underline{\quad}$

22) $\underline{\quad} - 8 = 10$

23) $\underline{\quad} - 7 = 8$

24) $\underline{\quad} - 9 = 11$

25) $\underline{\quad} - 11 = 1$

26) $20 - \underline{\quad} = 10$

27) $8 - \underline{\quad} = 2$

28) $\underline{\quad} - 2 = 6$

29) $\underline{\quad} - 10 = 5$

30) $14 - \underline{\quad} = 1$

31) $16 - \underline{\quad} = 9$

32) $15 - \underline{\quad} = 9$

33) $16 - \underline{\quad} = 13$

34) $\underline{\quad} - 2 = 8$

35) $\underline{\quad} - 7 = 3$

36) $15 - 12 = \underline{\quad}$

37) $\underline{\quad} - 7 = 3$

38) $\underline{\quad} - 3 = 10$

39) $\underline{\quad} - 10 = 8$

40) $20 - 11 = \underline{\quad}$

41) $18 - 3 = \underline{\quad}$

42) $14 - \underline{\quad} = 8$

43) $\underline{\quad} - 0 = 10$

44) $18 - \underline{\quad} = 5$

45) $\underline{\quad} - 11 = 7$

46) $\underline{\quad} - 11 = 2$

47) $\underline{\quad} - 7 = 6$

48) $\underline{\quad} - 3 = 5$

49) $14 - \underline{\quad} = 13$

50) $8 - \underline{\quad} = 7$

51) $\underline{\quad} - 9 = 7$

52) $\underline{\quad} - 6 = 12$

53) $\underline{\quad} - 2 = 13$

54) $\underline{\quad} - 12 = 2$

55) $19 - \underline{\quad} = 6$

56) $15 - \underline{\quad} = 8$

57) $\underline{\quad} - 4 = 15$

58) $\underline{\quad} - 1 = 5$

59) $10 - \underline{\quad} = 5$

60) $\underline{\quad} - 2 = 11$

Answer Key
Subtraction Missing Addends 0 - 20

1) 16 − **3** = 13

2) 15 − **13** = 2

3) 14 − **9** = 5

4) 19 − **9** = 10

5) 6 − **5** = 1

6) 18 − **3** = 15

7) 18 − **16** = 2

8) **15** − 4 = 11

9) **19** − 15 = 4

10) 12 − 7 = **5**

11) 19 − **9** = 10

12) 12 − **3** = 9

13) 9 − 5 = **4**

14) 14 − 10 = **4**

15) 12 − 4 = **8**

16) **19** − 14 = 5

17) 10 − 7 = **3**

18) **7** − 1 = 6

19) 13 − 12 = **1**

20) **8** − 0 = 8

21) 11 − 6 = **5**

22) **18** − 8 = 10

23) **15** − 7 = 8

24) **20** − 9 = 11

25) **12** − 11 = 1

26) 20 − **10** = 10

27) 8 − **6** = 2

28) **8** − 2 = 6

29) **15** − 10 = 5

30) 14 − **13** = 1

31) 16 − **7** = 9

32) 15 − **6** = 9

33) 16 − **3** = 13

34) **10** − 2 = 8

35) **10** − 7 = 3

36) 15 − 12 = **3**

37) **10** − 7 = 3

38) **13** − 3 = 10

39) **18** − 10 = 8

40) 20 − 11 = **9**

41) 18 − 3 = **15**

42) 14 − **6** = 8

43) **10** − 0 = 10

44) 18 − **13** = 5

45) **18** − 11 = 7

46) **13** − 11 = 2

47) **13** − 7 = 6

48) **8** − 3 = 5

49) 14 − **1** = 13

50) 8 − **1** = 7

51) **16** − 9 = 7

52) **18** − 6 = 12

53) **15** − 2 = 13

54) **14** − 12 = 2

55) 19 − **13** = 6

56) 15 − **7** = 8

57) **19** − 4 = 15

58) **6** − 1 = 5

59) 10 − **5** = 5

60) **13** − 2 = 11

Date : _________

Time : _________

Name : _________

Score : _____/60

Subtraction Missing Addends 0 - 20

1) $10 - \underline{\quad} = 6$

2) $17 - 1 = \underline{\quad}$

3) $\underline{\quad} - 2 = 18$

4) $\underline{\quad} - 2 = 12$

5) $\underline{\quad} - 5 = 11$

6) $19 - 7 = \underline{\quad}$

7) $12 - \underline{\quad} = 10$

8) $\underline{\quad} - 9 = 3$

9) $12 - 11 = \underline{\quad}$

10) $12 - 12 = \underline{\quad}$

11) $6 - \underline{\quad} = 5$

12) $11 - \underline{\quad} = 4$

13) $\underline{\quad} - 16 = 2$

14) $\underline{\quad} - 6 = 2$

15) $10 - \underline{\quad} = 4$

16) $16 - 15 = \underline{\quad}$

17) $16 - 5 = \underline{\quad}$

18) $18 - \underline{\quad} = 7$

19) $20 - \underline{\quad} = 11$

20) $18 - \underline{\quad} = 14$

21) $\underline{\quad} - 2 = 12$

22) $9 - \underline{\quad} = 7$

23) $17 - \underline{\quad} = 16$

24) $13 - \underline{\quad} = 11$

25) $19 - \underline{\quad} = 15$

26) $16 - \underline{\quad} = 9$

27) $\underline{\quad} - 9 = 10$

28) $\underline{\quad} - 5 = 15$

29) $17 - \underline{\quad} = 1$

30) $\underline{\quad} - 11 = 2$

31) $\underline{\quad} - 4 = 9$

32) $10 - 9 = \underline{\quad}$

33) $\underline{\quad} - 4 = 4$

34) $\underline{\quad} - 6 = 7$

35) $\underline{\quad} - 1 = 19$

36) $\underline{\quad} - 6 = 1$

37) $\underline{\quad} - 8 = 7$

38) $\underline{\quad} - 18 = 1$

39) $18 - \underline{\quad} = 0$

40) $\underline{\quad} - 5 = 7$

41) $\underline{\quad} - 1 = 11$

42) $20 - 8 = \underline{\quad}$

43) $13 - \underline{\quad} = 7$

44) $\underline{\quad} - 12 = 8$

45) $\underline{\quad} - 6 = 12$

46) $11 - \underline{\quad} = 2$

47) $5 - 1 = \underline{\quad}$

48) $\underline{\quad} - 0 = 17$

49) $19 - 17 = \underline{\quad}$

50) $11 - \underline{\quad} = 8$

51) $\underline{\quad} - 3 = 10$

52) $13 - 9 = \underline{\quad}$

53) $20 - \underline{\quad} = 3$

54) $\underline{\quad} - 8 = 9$

55) $16 - \underline{\quad} = 5$

56) $12 - \underline{\quad} = 6$

57) $\underline{\quad} - 9 = 6$

58) $\underline{\quad} - 1 = 1$

59) $\underline{\quad} - 7 = 8$

60) $10 - \underline{\quad} = 3$

Answer Key
Subtraction Missing Addends 0 - 20

1) $10 - \mathbf{4} = 6$

2) $17 - 1 = \mathbf{16}$

3) $\mathbf{20} - 2 = 18$

4) $\mathbf{14} - 2 = 12$

5) $\mathbf{16} - 5 = 11$

6) $19 - 7 = \mathbf{12}$

7) $12 - \mathbf{2} = 10$

8) $\mathbf{12} - 9 = 3$

9) $12 - 11 = \mathbf{1}$

10) $12 - 12 = \mathbf{0}$

11) $6 - \mathbf{1} = 5$

12) $11 - \mathbf{7} = 4$

13) $\mathbf{18} - 16 = 2$

14) $\mathbf{8} - 6 = 2$

15) $10 - \mathbf{6} = 4$

16) $16 - 15 = \mathbf{1}$

17) $16 - 5 = \mathbf{11}$

18) $18 - \mathbf{11} = 7$

19) $20 - \mathbf{9} = 11$

20) $18 - \mathbf{4} = 14$

21) $\mathbf{14} - 2 = 12$

22) $9 - \mathbf{2} = 7$

23) $17 - \mathbf{1} = 16$

24) $13 - \mathbf{2} = 11$

25) $19 - \mathbf{4} = 15$

26) $16 - \mathbf{7} = 9$

27) $\mathbf{19} - 9 = 10$

28) $\mathbf{20} - 5 = 15$

29) $17 - \mathbf{16} = 1$

30) $\mathbf{13} - 11 = 2$

31) $\mathbf{13} - 4 = 9$

32) $10 - 9 = \mathbf{1}$

33) $\mathbf{8} - 4 = 4$

34) $\mathbf{13} - 6 = 7$

35) $\mathbf{20} - 1 = 19$

36) $\mathbf{7} - 6 = 1$

37) $\mathbf{15} - 8 = 7$

38) $\mathbf{19} - 18 = 1$

39) $18 - \mathbf{18} = 0$

40) $\mathbf{12} - 5 = 7$

41) $\mathbf{12} - 1 = 11$

42) $20 - 8 = \mathbf{12}$

43) $13 - \mathbf{6} = 7$

44) $\mathbf{20} - 12 = 8$

45) $\mathbf{18} - 6 = 12$

46) $11 - \mathbf{9} = 2$

47) $5 - 1 = \mathbf{4}$

48) $\mathbf{17} - 0 = 17$

49) $19 - 17 = \mathbf{2}$

50) $11 - \mathbf{3} = 8$

51) $\mathbf{13} - 3 = 10$

52) $13 - 9 = \mathbf{4}$

53) $20 - \mathbf{17} = 3$

54) $\mathbf{17} - 8 = 9$

55) $16 - \mathbf{11} = 5$

56) $12 - \mathbf{6} = 6$

57) $\mathbf{15} - 9 = 6$

58) $\mathbf{2} - 1 = 1$

59) $\mathbf{15} - 7 = 8$

60) $10 - \mathbf{7} = 3$

Subtraction Missing Addends 0 - 20

1) _____ − 7 = 3   2) 16 − _____ = 5   3) _____ − 10 = 0

4) 20 − 1 = _____   5) 7 − _____ = 5   6) _____ − 7 = 6

7) 20 − 15 = _____   8) _____ − 8 = 7   9) 18 − 12 = _____

10) _____ − 1 = 2   11) 10 − 5 = _____   12) 16 − 1 = _____

13) _____ − 2 = 16   14) _____ − 16 = 2   15) _____ − 4 = 8

16) 5 − _____ = 1   17) _____ − 12 = 6   18) _____ − 17 = 1

19) 5 − _____ = 1   20) _____ − 11 = 1   21) _____ − 7 = 12

22) 18 − 1 = _____   23) 15 − 1 = _____   24) 7 − 1 = _____

25) 11 − _____ = 0   26) _____ − 5 = 8   27) _____ − 17 = 0

28) 12 − _____ = 1   29) _____ − 4 = 6   30) _____ − 6 = 2

31) 13 − _____ = 12   32) _____ − 16 = 3   33) 11 − 1 = _____

34) 12 − 5 = _____   35) 7 − _____ = 0   36) 19 − 7 = _____

37) _____ − 2 = 15   38) 20 − 15 = _____   39) 19 − _____ = 18

40) _____ − 5 = 12   41) 9 − 8 = _____   42) 16 − _____ = 0

43) _____ − 1 = 3   44) _____ − 5 = 3   45) _____ − 2 = 9

46) 20 − 20 = _____   47) _____ − 15 = 5   48) _____ − 2 = 13

49) 17 − _____ = 3   50) 17 − _____ = 9   51) _____ − 1 = 11

52) _____ − 15 = 1   53) _____ − 14 = 6   54) 14 − _____ = 13

55) 11 − _____ = 3   56) 19 − 1 = _____   57) 10 − 9 = _____

58) 15 − 9 = _____   59) 12 − 1 = _____   60) 3 − 2 = _____

Answer Key
Subtraction Missing Addends 0 - 20

1) **10** − 7 = 3

2) 16 − **11** = 5

3) **10** − 10 = 0

4) 20 − 1 = **19**

5) 7 − **2** = 5

6) **13** − 7 = 6

7) 20 − 15 = **5**

8) **15** − 8 = 7

9) 18 − 12 = **6**

10) **3** − 1 = 2

11) 10 − 5 = **5**

12) 16 − 1 = **15**

13) **18** − 2 = 16

14) **18** − 16 = 2

15) **12** − 4 = 8

16) 5 − **4** = 1

17) **18** − 12 = 6

18) **18** − 17 = 1

19) 5 − **4** = 1

20) **12** − 11 = 1

21) **19** − 7 = 12

22) 18 − 1 = **17**

23) 15 − 1 = **14**

24) 7 − 1 = **6**

25) 11 − **11** = 0

26) **13** − 5 = 8

27) **17** − 17 = 0

28) 12 − **11** = 1

29) **10** − 4 = 6

30) **8** − 6 = 2

31) 13 − **1** = 12

32) **19** − 16 = 3

33) 11 − 1 = **10**

34) 12 − 5 = **7**

35) 7 − **7** = 0

36) 19 − 7 = **12**

37) **17** − 2 = 15

38) 20 − 15 = **5**

39) 19 − **1** = 18

40) **17** − 5 = 12

41) 9 − 8 = **1**

42) 16 − **16** = 0

43) **4** − 1 = 3

44) **8** − 5 = 3

45) **11** − 2 = 9

46) 20 − 20 = **0**

47) **20** − 15 = 5

48) **15** − 2 = 13

49) 17 − **14** = 3

50) 17 − **8** = 9

51) **12** − 1 = 11

52) **16** − 15 = 1

53) **20** − 14 = 6

54) 14 − **1** = 13

55) 11 − **8** = 3

56) 19 − 1 = **18**

57) 10 − 9 = **1**

58) 15 − 9 = **6**

59) 12 − 1 = **11**

60) 3 − 2 = **1**

Subtraction Missing Addends 0 - 20

1) _____ − 4 = 5

2) 16 − _____ = 4

3) _____ − 18 = 2

4) _____ − 12 = 7

5) _____ − 16 = 4

6) 17 − _____ = 3

7) 11 − _____ = 1

8) _____ − 0 = 9

9) _____ − 1 = 13

10) _____ − 14 = 6

11) 20 − 17 = _____

12) 19 − _____ = 9

13) _____ − 7 = 9

14) 13 − _____ = 11

15) _____ − 5 = 14

16) 18 − _____ = 5

17) 11 − _____ = 4

18) _____ − 7 = 5

19) _____ − 6 = 7

20) 20 − 18 = _____

21) 13 − _____ = 0

22) 16 − _____ = 15

23) _____ − 9 = 6

24) 20 − _____ = 4

25) 17 − _____ = 3

26) _____ − 6 = 12

27) 18 − _____ = 9

28) 17 − 9 = _____

29) 14 − _____ = 0

30) _____ − 6 = 7

31) 20 − _____ = 19

32) _____ − 8 = 5

33) _____ − 2 = 2

34) 20 − 15 = _____

35) _____ − 4 = 5

36) 19 − 15 = _____

37) 20 − 11 = _____

38) 17 − _____ = 12

39) _____ − 9 = 7

40) _____ − 5 = 15

41) 15 − 8 = _____

42) 11 − 7 = _____

43) 5 − 1 = _____

44) _____ − 13 = 4

45) _____ − 15 = 3

46) _____ − 8 = 6

47) 16 − _____ = 1

48) 17 − _____ = 4

49) 16 − _____ = 11

50) _____ − 5 = 9

51) _____ − 13 = 2

52) 12 − 6 = _____

53) 10 − _____ = 8

54) 9 − 7 = _____

55) 11 − 3 = _____

56) _____ − 6 = 0

57) _____ − 7 = 3

58) 20 − _____ = 9

59) 7 − _____ = 0

60) 2 − _____ = 1

1) **9** – 4 = 5

2) 16 – **12** = 4

3) **20** – 18 = 2

4) **19** – 12 = 7

5) **20** – 16 = 4

6) 17 – **14** = 3

7) 11 – **10** = 1

8) **9** – 0 = 9

9) **14** – 1 = 13

10) **20** – 14 = 6

11) 20 – 17 = **3**

12) 19 – **10** = 9

13) **16** – 7 = 9

14) 13 – **2** = 11

15) **19** – 5 = 14

16) 18 – **13** = 5

17) 11 – **7** = 4

18) **12** – 7 = 5

19) **13** – 6 = 7

20) 20 – 18 = **2**

21) 13 – **13** = 0

22) 16 – **1** = 15

23) **15** – 9 = 6

24) 20 – **16** = 4

25) 17 – **14** = 3

26) **18** – 6 = 12

27) 18 – **9** = 9

28) 17 – 9 = **8**

29) 14 – **14** = 0

30) **13** – 6 = 7

31) 20 – **1** = 19

32) **13** – 8 = 5

33) **4** – 2 = 2

34) 20 – 15 = **5**

35) **9** – 4 = 5

36) 19 – 15 = **4**

37) 20 – 11 = **9**

38) 17 – **5** = 12

39) **16** – 9 = 7

40) **20** – 5 = 15

41) 15 – 8 = **7**

42) 11 – 7 = **4**

43) 5 – 1 = **4**

44) **17** – 13 = 4

45) **18** – 15 = 3

46) **14** – 8 = 6

47) 16 – **15** = 1

48) 17 – **13** = 4

49) 16 – **5** = 11

50) **14** – 5 = 9

51) **15** – 13 = 2

52) 12 – 6 = **6**

53) 10 – **2** = 8

54) 9 – 7 = **2**

55) 11 – 3 = **8**

56) **6** – 6 = 0

57) **10** – 7 = 3

58) 20 – **11** = 9

59) 7 – **7** = 0

60) 2 – **1** = 1

Subtraction Missing Addends 0 - 20

1) _____ – 9 = 11

2) 10 – 1 = _____

3) 7 – 2 = _____

4) _____ – 5 = 14

5) _____ – 3 = 14

6) 15 – 4 = _____

7) _____ – 6 = 11

8) 18 – _____ = 3

9) _____ – 10 = 7

10) _____ – 1 = 5

11) 15 – 4 = _____

12) _____ – 13 = 3

13) _____ – 14 = 2

14) 20 – _____ = 15

15) 17 – _____ = 10

16) 17 – _____ = 1

17) 10 – _____ = 3

18) 5 – 3 = _____

19) _____ – 11 = 2

20) _____ – 11 = 7

21) 10 – _____ = 9

22) 10 – _____ = 9

23) 19 – _____ = 17

24) 17 – 9 = _____

25) 18 – 15 = _____

26) 20 – 17 = _____

27) 10 – _____ = 4

28) 14 – _____ = 2

29) _____ – 4 = 14

30) 5 – _____ = 4

31) 18 – _____ = 7

32) _____ – 9 = 11

33) 17 – _____ = 14

34) _____ – 4 = 5

35) 17 – _____ = 7

36) _____ – 2 = 2

37) 16 – 4 = _____

38) 10 – _____ = 1

39) 10 – _____ = 0

40) _____ – 18 = 2

41) 8 – _____ = 2

42) 16 – _____ = 12

43) _____ – 14 = 3

44) 20 – _____ = 17

45) _____ – 8 = 5

46) _____ – 14 = 1

47) 6 – _____ = 3

48) 8 – _____ = 2

49) _____ – 11 = 3

50) 16 – _____ = 4

51) _____ – 4 = 13

52) 8 – 2 = _____

53) 15 – _____ = 12

54) 10 – _____ = 6

55) _____ – 3 = 1

56) _____ – 6 = 8

57) 10 – _____ = 7

58) _____ – 5 = 5

59) _____ – 7 = 6

60) _____ – 17 = 0

Answer Key

Subtraction Missing Addends 0 - 20

1) **20** − 9 = 11

2) 10 − 1 = **9**

3) 7 − 2 = **5**

4) **19** − 5 = 14

5) **17** − 3 = 14

6) 15 − 4 = **11**

7) **17** − 6 = 11

8) 18 − **15** = 3

9) **17** − 10 = 7

10) **6** − 1 = 5

11) 15 − 4 = **11**

12) **16** − 13 = 3

13) **16** − 14 = 2

14) 20 − **5** = 15

15) 17 − **7** = 10

16) 17 − **16** = 1

17) 10 − **7** = 3

18) 5 − 3 = **2**

19) **13** − 11 = 2

20) **18** − 11 = 7

21) 10 − **1** = 9

22) 10 − **1** = 9

23) 19 − **2** = 17

24) 17 − 9 = **8**

25) 18 − 15 = **3**

26) 20 − 17 = **3**

27) 10 − **6** = 4

28) 14 − **12** = 2

29) **18** − 4 = 14

30) 5 − **1** = 4

31) 18 − **11** = 7

32) **20** − 9 = 11

33) 17 − **3** = 14

34) **9** − 4 = 5

35) 17 − **10** = 7

36) **4** − 2 = 2

37) 16 − 4 = **12**

38) 10 − **9** = 1

39) 10 − **10** = 0

40) **20** − 18 = 2

41) 8 − **6** = 2

42) 16 − **4** = 12

43) **17** − 14 = 3

44) 20 − **3** = 17

45) **13** − 8 = 5

46) **15** − 14 = 1

47) 6 − **3** = 3

48) 8 − **6** = 2

49) **14** − 11 = 3

50) 16 − **12** = 4

51) **17** − 4 = 13

52) 8 − 2 = **6**

53) 15 − **3** = 12

54) 10 − **4** = 6

55) **4** − 3 = 1

56) **14** − 6 = 8

57) 10 − **3** = 7

58) **10** − 5 = 5

59) **13** − 7 = 6

60) **17** − 17 = 0

Date : _________   Name : __________
Time : _________   Score : _____/60

 Subtraction Missing Addends 0 - 20 

1) 19 – _____ = 3   2) 15 – _____ = 9   3) _____ – 2 = 17

4) _____ – 5 = 15   5) 11 – 10 = _____   6) 15 – 5 = _____

7) _____ – 2 = 3   8) _____ – 7 = 12   9) 3 – _____ = 0

10) 15 – 11 = _____   11) 8 – _____ = 3   12) 11 – _____ = 2

13) 9 – _____ = 5   14) 15 – _____ = 1   15) _____ – 3 = 8

16) 14 – _____ = 4   17) 17 – 10 = _____   18) _____ – 3 = 7

19) 17 – _____ = 4   20) _____ – 13 = 2   21) _____ – 2 = 11

22) 12 – _____ = 10   23) 17 – _____ = 3   24) _____ – 3 = 3

25) 14 – 2 = _____   26) 16 – _____ = 5   27) 17 – _____ = 7

28) _____ – 9 = 1   29) _____ – 1 = 6   30) 15 – _____ = 8

31) 9 – 2 = _____   32) 8 – _____ = 4   33) _____ – 11 = 1

34) 14 – _____ = 0   35) _____ – 3 = 0   36) _____ – 9 = 3

37) 9 – 2 = _____   38) 9 – _____ = 3   39) 20 – _____ = 17

40) 16 – _____ = 9   41) 10 – _____ = 4   42) 16 – 14 = _____

43) 19 – _____ = 7   44) 13 – _____ = 3   45) 15 – 6 = _____

46) _____ – 1 = 15   47) _____ – 10 = 10   48) _____ – 4 = 4

49) _____ – 5 = 5   50) 20 – _____ = 5   51) 14 – _____ = 9

52) 12 – _____ = 9   53) 19 – _____ = 8   54) 11 – _____ = 2

55) 16 – _____ = 2   56) 16 – _____ = 6   57) _____ – 3 = 1

58) 5 – _____ = 2   59) 12 – _____ = 8   60) _____ – 13 = 2

Answer Key
Subtraction Missing Addends 0 - 20

1) $19 - \mathbf{16} = 3$

2) $15 - \mathbf{6} = 9$

3) $\mathbf{19} - 2 = 17$

4) $\mathbf{20} - 5 = 15$

5) $11 - 10 = \mathbf{1}$

6) $15 - 5 = \mathbf{10}$

7) $\mathbf{5} - 2 = 3$

8) $\mathbf{19} - 7 = 12$

9) $3 - \mathbf{3} = 0$

10) $15 - 11 = \mathbf{4}$

11) $8 - \mathbf{5} = 3$

12) $11 - \mathbf{9} = 2$

13) $9 - \mathbf{4} = 5$

14) $15 - \mathbf{14} = 1$

15) $\mathbf{11} - 3 = 8$

16) $14 - \mathbf{10} = 4$

17) $17 - 10 = \mathbf{7}$

18) $\mathbf{10} - 3 = 7$

19) $17 - \mathbf{13} = 4$

20) $\mathbf{15} - 13 = 2$

21) $\mathbf{13} - 2 = 11$

22) $12 - \mathbf{2} = 10$

23) $17 - \mathbf{14} = 3$

24) $\mathbf{6} - 3 = 3$

25) $14 - 2 = \mathbf{12}$

26) $16 - \mathbf{11} = 5$

27) $17 - \mathbf{10} = 7$

28) $\mathbf{10} - 9 = 1$

29) $\mathbf{7} - 1 = 6$

30) $15 - \mathbf{7} = 8$

31) $9 - 2 = \mathbf{7}$

32) $8 - \mathbf{4} = 4$

33) $\mathbf{12} - 11 = 1$

34) $14 - \mathbf{14} = 0$

35) $\mathbf{3} - 3 = 0$

36) $\mathbf{12} - 9 = 3$

37) $9 - 2 = \mathbf{7}$

38) $9 - \mathbf{6} = 3$

39) $20 - \mathbf{3} = 17$

40) $16 - \mathbf{7} = 9$

41) $10 - \mathbf{6} = 4$

42) $16 - 14 = \mathbf{2}$

43) $19 - \mathbf{12} = 7$

44) $13 - \mathbf{10} = 3$

45) $15 - 6 = \mathbf{9}$

46) $\mathbf{16} - 1 = 15$

47) $\mathbf{20} - 10 = 10$

48) $\mathbf{8} - 4 = 4$

49) $\mathbf{10} - 5 = 5$

50) $20 - \mathbf{15} = 5$

51) $14 - \mathbf{5} = 9$

52) $12 - \mathbf{3} = 9$

53) $19 - \mathbf{11} = 8$

54) $11 - \mathbf{9} = 2$

55) $16 - \mathbf{14} = 2$

56) $16 - \mathbf{10} = 6$

57) $\mathbf{4} - 3 = 1$

58) $5 - \mathbf{3} = 2$

59) $12 - \mathbf{4} = 8$

60) $\mathbf{15} - 13 = 2$

Mixed Problems 0 - 20

1) 2 + 18 = _____   2) 20 + 0 = _____   3) 10 − 6 = _____

4) 14 − 1 = _____   5) 14 − 9 = _____   6) 19 − 3 = _____

7) 3 + 12 = _____   8) 4 + 11 = _____   9) 4 + 16 = _____

10) 10 + 5 = _____   11) 15 − 8 = _____   12) 5 − 4 = _____

13) 9 − 7 = _____   14) 0 + 16 = _____   15) 19 − 6 = _____

16) 7 + 4 = _____   17) 19 − 7 = _____   18) 1 − 0 = _____

19) 12 + 4 = _____   20) 2 + 8 = _____   21) 7 − 4 = _____

22) 3 + 17 = _____   23) 4 − 1 = _____   24) 0 + 20 = _____

25) 13 − 8 = _____   26) 17 − 13 = _____   27) 4 + 8 = _____

28) 7 + 5 = _____   29) 8 − 5 = _____   30) 16 + 1 = _____

31) 8 − 3 = _____   32) 2 + 3 = _____   33) 6 + 5 = _____

34) 20 − 8 = _____   35) 20 + 0 = _____   36) 13 − 8 = _____

37) 15 − 12 = _____   38) 17 + 4 = _____   39) 10 + 8 = _____

40) 9 − 3 = _____   41) 1 + 17 = _____   42) 10 + 9 = _____

43) 2 − 1 = _____   44) 16 − 10 = _____   45) 18 − 5 = _____

46) 11 + 6 = _____   47) 14 + 1 = _____   48) 19 − 10 = _____

49) 17 − 3 = _____   50) 15 − 10 = _____   51) 5 + 11 = _____

52) 0 + 20 = _____   53) 0 + 13 = _____   54) 15 − 13 = _____

55) 12 − 4 = _____   56) 3 + 17 = _____   57) 12 − 6 = _____

58) 13 + 7 = _____   59) 18 − 15 = _____   60) 2 + 17 = _____

Answer Key
Mixed Problems 0 - 20

1) 2 + 18 = **20**

2) 20 + 0 = **20**

3) 10 − 6 = **4**

4) 14 − 1 = **13**

5) 14 − 9 = **5**

6) 19 − 3 = **16**

7) 3 + 12 = **15**

8) 4 + 11 = **15**

9) 4 + 16 = **20**

10) 10 + 5 = **15**

11) 15 − 8 = **7**

12) 5 − 4 = **1**

13) 9 − 7 = **2**

14) 0 + 16 = **16**

15) 19 − 6 = **13**

16) 7 + 4 = **11**

17) 19 − 7 = **12**

18) 1 − 0 = **1**

19) 12 + 4 = **16**

20) 2 + 8 = **10**

21) 7 − 4 = **3**

22) 3 + 17 = **20**

23) 4 − 1 = **3**

24) 0 + 20 = **20**

25) 13 − 8 = **5**

26) 17 − 13 = **4**

27) 4 + 8 = **12**

28) 7 + 5 = **12**

29) 8 − 5 = **3**

30) 16 + 1 = **17**

31) 8 − 3 = **5**

32) 2 + 3 = **5**

33) 6 + 5 = **11**

34) 20 − 8 = **12**

35) 20 + 0 = **20**

36) 13 − 8 = **5**

37) 15 − 12 = **3**

38) 17 + 4 = **21**

39) 10 + 8 = **18**

40) 9 − 3 = **6**

41) 1 + 17 = **18**

42) 10 + 9 = **19**

43) 2 − 1 = **1**

44) 16 − 10 = **6**

45) 18 − 5 = **13**

46) 11 + 6 = **17**

47) 14 + 1 = **15**

48) 19 − 10 = **9**

49) 17 − 3 = **14**

50) 15 − 10 = **5**

51) 5 + 11 = **16**

52) 0 + 20 = **20**

53) 0 + 13 = **13**

54) 15 − 13 = **2**

55) 12 − 4 = **8**

56) 3 + 17 = **20**

57) 12 − 6 = **6**

58) 13 + 7 = **20**

59) 18 − 15 = **3**

60) 2 + 17 = **19**

Mixed Problems 0 - 20

1) _____ + 9 = 19

2) _____ − 5 = 8

3) 15 − 3 = _____

4) 0 + _____ = 7

5) _____ + 10 = 11

6) 10 − _____ = 5

7) 11 + 8 = _____

8) 19 − _____ = 8

9) 13 + _____ = 14

10) _____ − 1 = 11

11) 1 + _____ = 15

12) 13 − _____ = 6

13) 11 − _____ = 1

14) _____ + 1 = 16

15) 7 − 6 = _____

16) _____ + 1 = 20

17) _____ + 0 = 18

18) _____ + 16 = 18

19) 15 − _____ = 14

20) 16 − _____ = 7

21) 2 + _____ = 16

22) _____ − 10 = 10

23) 8 + _____ = 20

24) 20 − _____ = 4

25) 19 + _____ = 20

26) _____ − 15 = 2

27) _____ + 2 = 11

28) _____ − 0 = 11

29) _____ − 14 = 6

30) _____ − 2 = 0

31) _____ + 0 = 20

32) 17 + _____ = 19

33) 17 + _____ = 18

34) _____ − 14 = 1

35) _____ + 0 = 6

36) 13 − 6 = _____

37) 10 + _____ = 19

38) 19 − _____ = 3

39) 20 − _____ = 9

40) _____ + 17 = 20

41) 15 − _____ = 6

42) _____ − 3 = 14

43) _____ + 6 = 16

44) _____ + 15 = 20

45) _____ + 2 = 16

46) 1 + _____ = 19

47) 17 − _____ = 17

48) _____ − 11 = 0

49) _____ − 16 = 3

50) 4 + _____ = 14

51) 12 − 10 = _____

52) 18 + 0 = _____

53) _____ − 16 = 3

54) _____ − 8 = 11

55) _____ + 16 = 19

56) _____ + 4 = 15

57) 2 + _____ = 17

58) _____ − 5 = 14

59) _____ − 5 = 3

60) _____ + 7 = 14

Answer Key
Mixed Problems 0 - 20

1) **10** + 9 = 19

2) **13** − 5 = 8

3) 15 − 3 = **12**

4) 0 + **7** = 7

5) **1** + 10 = 11

6) 10 − **5** = 5

7) 11 + 8 = **19**

8) 19 − **11** = 8

9) 13 + **1** = 14

10) **12** − 1 = 11

11) 1 + **14** = 15

12) 13 − **7** = 6

13) 11 − **10** = 1

14) **15** + 1 = 16

15) 7 − 6 = **1**

16) **19** + 1 = 20

17) **18** + 0 = 18

18) **2** + 16 = 18

19) 15 − **1** = 14

20) 16 − **9** = 7

21) 2 + **14** = 16

22) **20** − 10 = 10

23) 8 + **12** = 20

24) 20 − **16** = 4

25) 19 + **1** = 20

26) **17** − 15 = 2

27) **9** + 2 = 11

28) **11** − 0 = 11

29) **20** − 14 = 6

30) **2** − 2 = 0

31) **20** + 0 = 20

32) 17 + **2** = 19

33) 17 + **1** = 18

34) **15** − 14 = 1

35) **6** + 0 = 6

36) 13 − 6 = **7**

37) 10 + **9** = 19

38) 19 − **16** = 3

39) 20 − **11** = 9

40) **3** + 17 = 20

41) 15 − **9** = 6

42) **17** − 3 = 14

43) **10** + 6 = 16

44) **5** + 15 = 20

45) **14** + 2 = 16

46) 1 + **18** = 19

47) 17 − **0** = 17

48) **11** − 11 = 0

49) **19** − 16 = 3

50) 4 + **10** = 14

51) 12 − 10 = **2**

52) 18 + 0 = **18**

53) **19** − 16 = 3

54) **19** − 8 = 11

55) **3** + 16 = 19

56) **11** + 4 = 15

57) 2 + **15** = 17

58) **19** − 5 = 14

59) **8** − 5 = 3

60) **7** + 7 = 14

Date : _________

Time : _________

Name : _________

Score : _____/60

Mixed Problems 0 - 20

1) 9 – 4 = _____

2) 10 + 6 = _____

3) 7 – 6 = _____

4) 19 + 0 = _____

5) 8 + 6 = _____

6) 20 – 3 = _____

7) 9 + 11 = _____

8) 4 – 3 = _____

9) 1 + 12 = _____

10) 18 – 17 = _____

11) 6 + 13 = _____

12) 17 – 14 = _____

13) 14 – 6 = _____

14) 16 – 6 = _____

15) 17 + 1 = _____

16) 17 + 2 = _____

17) 11 + 2 = _____

18) 15 – 6 = _____

19) 16 – 12 = _____

20) 17 + 1 = _____

21) 9 + 8 = _____

22) 8 + 9 = _____

23) 20 – 7 = _____

24) 8 – 8 = _____

25) 18 – 13 = _____

26) 1 + 9 = _____

27) 1 + 18 = _____

28) 13 – 3 = _____

29) 16 – 7 = _____

30) 6 – 4 = _____

31) 4 + 6 = _____

32) 16 + 1 = _____

33) 20 – 1 = _____

34) 10 – 9 = _____

35) 5 + 9 = _____

36) 0 + 20 = _____

37) 19 – 12 = _____

38) 12 + 3 = _____

39) 8 + 9 = _____

40) 20 – 10 = _____

41) 19 – 1 = _____

42) 12 + 7 = _____

43) 5 – 0 = _____

44) 13 + 0 = _____

45) 8 + 3 = _____

46) 14 – 13 = _____

47) 15 + 4 = _____

48) 13 – 5 = _____

49) 10 – 6 = _____

50) 6 + 13 = _____

51) 8 – 6 = _____

52) 1 + 1 = _____

53) 14 + 6 = _____

54) 6 – 4 = _____

55) 19 – 18 = _____

56) 17 + 2 = _____

57) 5 + 8 = _____

58) 15 + 0 = _____

59) 6 – 3 = _____

60) 17 – 6 = _____

Answer Key
Mixed Problems 0 - 20

1) $9 - 4 = \mathbf{5}$

2) $10 + 6 = \mathbf{16}$

3) $7 - 6 = \mathbf{1}$

4) $19 + 0 = \mathbf{19}$

5) $8 + 6 = \mathbf{14}$

6) $20 - 3 = \mathbf{17}$

7) $9 + 11 = \mathbf{20}$

8) $4 - 3 = \mathbf{1}$

9) $1 + 12 = \mathbf{13}$

10) $18 - 17 = \mathbf{1}$

11) $6 + 13 = \mathbf{19}$

12) $17 - 14 = \mathbf{3}$

13) $14 - 6 = \mathbf{8}$

14) $16 - 6 = \mathbf{10}$

15) $17 + 1 = \mathbf{18}$

16) $17 + 2 = \mathbf{19}$

17) $11 + 2 = \mathbf{13}$

18) $15 - 6 = \mathbf{9}$

19) $16 - 12 = \mathbf{4}$

20) $17 + 1 = \mathbf{18}$

21) $9 + 8 = \mathbf{17}$

22) $8 + 9 = \mathbf{17}$

23) $20 - 7 = \mathbf{13}$

24) $8 - 8 = \mathbf{0}$

25) $18 - 13 = \mathbf{5}$

26) $1 + 9 = \mathbf{10}$

27) $1 + 18 = \mathbf{19}$

28) $13 - 3 = \mathbf{10}$

29) $16 - 7 = \mathbf{9}$

30) $6 - 4 = \mathbf{2}$

31) $4 + 6 = \mathbf{10}$

32) $16 + 1 = \mathbf{17}$

33) $20 - 1 = \mathbf{19}$

34) $10 - 9 = \mathbf{1}$

35) $5 + 9 = \mathbf{14}$

36) $0 + 20 = \mathbf{20}$

37) $19 - 12 = \mathbf{7}$

38) $12 + 3 = \mathbf{15}$

39) $8 + 9 = \mathbf{17}$

40) $20 - 10 = \mathbf{10}$

41) $19 - 1 = \mathbf{18}$

42) $12 + 7 = \mathbf{19}$

43) $5 - 0 = \mathbf{5}$

44) $13 + 0 = \mathbf{13}$

45) $8 + 3 = \mathbf{11}$

46) $14 - 13 = \mathbf{1}$

47) $15 + 4 = \mathbf{19}$

48) $13 - 5 = \mathbf{8}$

49) $10 - 6 = \mathbf{4}$

50) $6 + 13 = \mathbf{19}$

51) $8 - 6 = \mathbf{2}$

52) $1 + 1 = \mathbf{2}$

53) $14 + 6 = \mathbf{20}$

54) $6 - 4 = \mathbf{2}$

55) $19 - 18 = \mathbf{1}$

56) $17 + 2 = \mathbf{19}$

57) $5 + 8 = \mathbf{13}$

58) $15 + 0 = \mathbf{15}$

59) $6 - 3 = \mathbf{3}$

60) $17 - 6 = \mathbf{11}$

Date : _________

Name : _________

Time : _________

Score : ______/60

Mixed Problems 0 - 20

1) 16 +_____ = 17

2) _____ − 6 = 6

3) _____ − 5 = 12

4) 3 +_____ = 18

5) 12 +_____ = 12

6) _____ − 2 = 2

7) _____ + 14 = 15

8) _____ − 15 = 2

9) 1 + 3 = _____

10) 13 −_____ = 11

11) 1 +_____ = 11

12) _____ − 12 = 3

13) 13 −_____ = 12

14) 2 + 8 = _____

15) _____ − 6 = 7

16) 2 + 9 = _____

17) 9 +_____ = 19

18) 18 −_____ = 10

19) 9 +_____ = 11

20) _____ − 4 = 8

21) _____ + 3 = 11

22) 6 − 0 = _____

23) _____ − 14 = 1

24) 1 +_____ = 10

25) 13 −_____ = 1

26) 19 +_____ = 20

27) _____ + 15 = 16

28) 10 −_____ = 2

29) _____ − 12 = 8

30) 2 +_____ = 8

31) _____ − 4 = 2

32) _____ + 7 = 19

33) 4 + 6 = _____

34) 12 − 4 = _____

35) _____ − 5 = 6

36) 17 + 1 = _____

37) 2 + 17 = _____

38) 20 −_____ = 6

39) _____ + 7 = 16

40) 17 −_____ = 6

41) 2 +_____ = 4

42) 2 +_____ = 19

43) 19 − 1 = _____

44) _____ − 4 = 8

45) _____ + 1 = 15

46) 13 −_____ = 5

47) _____ − 13 = 2

48) 17 +_____ = 19

49) 12 +_____ = 17

50) 10 −_____ = 6

51) 8 + 6 = _____

52) 3 − 2 = _____

53) 2 + 5 = _____

54) _____ − 12 = 3

55) _____ + 6 = 14

56) _____ − 12 = 7

57) _____ − 11 = 2

58) _____ + 18 = 20

59) 14 + 5 = _____

60) 16 − 11 = _____

Answer Key

Mixed Problems 0 - 20

1) $16 + \mathbf{1} = 17$

2) $\mathbf{12} - 6 = 6$

3) $\mathbf{17} - 5 = 12$

4) $3 + \mathbf{15} = 18$

5) $12 + \mathbf{0} = 12$

6) $\mathbf{4} - 2 = 2$

7) $\mathbf{1} + 14 = 15$

8) $\mathbf{17} - 15 = 2$

9) $1 + 3 = \mathbf{4}$

10) $13 - \mathbf{2} = 11$

11) $1 + \mathbf{10} = 11$

12) $\mathbf{15} - 12 = 3$

13) $13 - \mathbf{1} = 12$

14) $2 + 8 = \mathbf{10}$

15) $\mathbf{13} - 6 = 7$

16) $2 + 9 = \mathbf{11}$

17) $9 + \mathbf{10} = 19$

18) $18 - \mathbf{8} = 10$

19) $9 + \mathbf{2} = 11$

20) $\mathbf{12} - 4 = 8$

21) $\mathbf{8} + 3 = 11$

22) $6 - 0 = \mathbf{6}$

23) $\mathbf{15} - 14 = 1$

24) $1 + \mathbf{9} = 10$

25) $13 - \mathbf{12} = 1$

26) $19 + \mathbf{1} = 20$

27) $\mathbf{1} + 15 = 16$

28) $10 - \mathbf{8} = 2$

29) $\mathbf{20} - 12 = 8$

30) $2 + \mathbf{6} = 8$

31) $\mathbf{6} - 4 = 2$

32) $\mathbf{12} + 7 = 19$

33) $4 + 6 = \mathbf{10}$

34) $12 - 4 = \mathbf{8}$

35) $\mathbf{11} - 5 = 6$

36) $17 + 1 = \mathbf{18}$

37) $2 + 17 = \mathbf{19}$

38) $20 - \mathbf{14} = 6$

39) $\mathbf{9} + 7 = 16$

40) $17 - \mathbf{11} = 6$

41) $2 + \mathbf{2} = 4$

42) $2 + \mathbf{17} = 19$

43) $19 - 1 = \mathbf{18}$

44) $\mathbf{12} - 4 = 8$

45) $\mathbf{14} + 1 = 15$

46) $13 - \mathbf{8} = 5$

47) $\mathbf{15} - 13 = 2$

48) $17 + \mathbf{2} = 19$

49) $12 + \mathbf{5} = 17$

50) $10 - \mathbf{4} = 6$

51) $8 + 6 = \mathbf{14}$

52) $3 - 2 = \mathbf{1}$

53) $2 + 5 = \mathbf{7}$

54) $\mathbf{15} - 12 = 3$

55) $\mathbf{8} + 6 = 14$

56) $\mathbf{19} - 12 = 7$

57) $\mathbf{13} - 11 = 2$

58) $\mathbf{2} + 18 = 20$

59) $14 + 5 = \mathbf{19}$

60) $16 - 11 = \mathbf{5}$

Date : _________   Name : _________

Time : _________   Score : _____/60

Mixed Problems 0 - 20

1) 4 − 1 = _____	2) 16 + 1 = _____	3) 18 − 13 = _____
4) 19 + 1 = _____	5) 19 − 9 = _____	6) 2 + 15 = _____
7) 8 + 3 = _____	8) 7 − 1 = _____	9) 12 − 4 = _____
10) 13 − 2 = _____	11) 8 + 7 = _____	12) 17 + 2 = _____
13) 1 + 19 = _____	14) 19 − 15 = _____	15) 16 − 10 = _____
16) 13 + 2 = _____	17) 11 − 6 = _____	18) 7 + 1 = _____
19) 20 − 14 = _____	20) 10 + 6 = _____	21) 2 + 16 = _____
22) 20 − 14 = _____	23) 1 + 16 = _____	24) 14 − 10 = _____
25) 3 + 13 = _____	26) 19 − 8 = _____	27) 13 − 6 = _____
28) 9 + 9 = _____	29) 2 + 18 = _____	30) 4 + 14 = _____
31) 16 − 9 = _____	32) 15 − 7 = _____	33) 13 + 6 = _____
34) 2 − 2 = _____	35) 16 − 3 = _____	36) 7 + 4 = _____
37) 9 + 3 = _____	38) 19 − 12 = _____	39) 3 + 11 = _____
40) 13 − 3 = _____	41) 6 + 4 = _____	42) 2 + 17 = _____
43) 8 − 7 = _____	44) 17 − 6 = _____	45) 12 − 8 = _____
46) 20 + 0 = _____	47) 19 + 1 = _____	48) 18 − 17 = _____
49) 15 − 5 = _____	50) 15 − 5 = _____	51) 18 + 0 = _____
52) 9 + 2 = _____	53) 13 − 10 = _____	54) 12 − 8 = _____
55) 1 + 18 = _____	56) 1 + 10 = _____	57) 11 − 8 = _____
58) 15 − 3 = _____	59) 2 + 17 = _____	60) 1 + 18 = _____

Answer Key

Mixed Problems 0 - 20

1) 4 − 1 = **3**

2) 16 + 1 = **17**

3) 18 − 13 = **5**

4) 19 + 1 = **20**

5) 19 − 9 = **10**

6) 2 + 15 = **17**

7) 8 + 3 = **11**

8) 7 − 1 = **6**

9) 12 − 4 = **8**

10) 13 − 2 = **11**

11) 8 + 7 = **15**

12) 17 + 2 = **19**

13) 1 + 19 = **20**

14) 19 − 15 = **4**

15) 16 − 10 = **6**

16) 13 + 2 = **15**

17) 11 − 6 = **5**

18) 7 + 1 = **8**

19) 20 − 14 = **6**

20) 10 + 6 = **16**

21) 2 + 16 = **18**

22) 20 − 14 = **6**

23) 1 + 16 = **17**

24) 14 − 10 = **4**

25) 3 + 13 = **16**

26) 19 − 8 = **11**

27) 13 − 6 = **7**

28) 9 + 9 = **18**

29) 2 + 18 = **20**

30) 4 + 14 = **18**

31) 16 − 9 = **7**

32) 15 − 7 = **8**

33) 13 + 6 = **19**

34) 2 − 2 = **0**

35) 16 − 3 = **13**

36) 7 + 4 = **11**

37) 9 + 3 = **12**

38) 19 − 12 = **7**

39) 3 + 11 = **14**

40) 13 − 3 = **10**

41) 6 + 4 = **10**

42) 2 + 17 = **19**

43) 8 − 7 = **1**

44) 17 − 6 = **11**

45) 12 − 8 = **4**

46) 20 + 0 = **20**

47) 19 + 1 = **20**

48) 18 − 17 = **1**

49) 15 − 5 = **10**

50) 15 − 5 = **10**

51) 18 + 0 = **18**

52) 9 + 2 = **11**

53) 13 − 10 = **3**

54) 12 − 8 = **4**

55) 1 + 18 = **19**

56) 1 + 10 = **11**

57) 11 − 8 = **3**

58) 15 − 3 = **12**

59) 2 + 17 = **19**

60) 1 + 18 = **19**

Date : _________   Name : __________

Time : _________   Score : _____/60

   Mixed Problems 0 - 20   

1) 4 + 4 = _____	2) 13 +_____= 19	3) 18 –_____= 12
4) _____ – 18 = 0	5) _____+ 7 = 14	6) 14 –_____= 4
7) 12 –_____= 9	8) 2 +_____= 20	9) _____+ 8 = 8
10) 14 –_____= 7	11) 9 + 6 = _____	12) 15 –_____= 14
13) 19 –_____= 16	14) 16 –_____= 10	15) 5 +_____= 7
16) 14 +_____= 17	17) _____+ 6 = 13	18) 15 – 8 = _____
19) 10 –_____= 9	20) _____+ 2 = 16	21) 17 –_____= 2
22) 17 +_____= 20	23) _____– 3 = 7	24) 6 +_____= 15
25) _____– 7 = 7	26) _____+ 9 = 17	27) 19 –_____= 6
28) 18 + 2 = _____	29) 17 +_____= 19	30) _____+ 18 = 20
31) 19 –_____= 8	32) 11 – 4 = _____	33) 7 + 10 = _____
34) 8 +_____= 13	35) 18 – 12 = _____	36) 19 – 6 = _____
37) _____+ 0 = 19	38) 18 – 2 = _____	39) 7 –_____= 6
40) 1 + 13 = _____	41) _____– 2 = 12	42) 15 + 4 = _____
43) _____+ 15 = 17	44) 15 – 15 = _____	45) _____+ 16 = 19
46) _____– 8 = 10	47) 10 + 4 = _____	48) _____– 5 = 11
49) 20 –_____= 14	50) 10 + 10 = _____	51) 20 +_____= 20
52) 11 –_____= 2	53) 17 –_____= 11	54) 15 –_____= 10
55) 2 +_____= 4	56) 5 + 13 = _____	57) _____– 0 = 3
58) 17 +_____= 18	59) 6 –_____= 2	60) 15 +_____= 19

Answer Key
Mixed Problems 0 - 20

1) 4 + 4 = **8**

2) 13 + **6** = 19

3) 18 − **6** = 12

4) **18** − 18 = 0

5) **7** + 7 = 14

6) 14 − **10** = 4

7) 12 − **3** = 9

8) 2 + **18** = 20

9) **0** + 8 = 8

10) 14 − **7** = 7

11) 9 + 6 = **15**

12) 15 − **1** = 14

13) 19 − **3** = 16

14) 16 − **6** = 10

15) 5 + **2** = 7

16) 14 + **3** = 17

17) **7** + 6 = 13

18) 15 − 8 = **7**

19) 10 − **1** = 9

20) **14** + 2 = 16

21) 17 − **15** = 2

22) 17 + **3** = 20

23) **10** − 3 = 7

24) 6 + **9** = 15

25) **14** − 7 = 7

26) **8** + 9 = 17

27) 19 − **13** = 6

28) 18 + 2 = **20**

29) 17 + **2** = 19

30) **2** + 18 = 20

31) 19 − **11** = 8

32) 11 − 4 = **7**

33) 7 + 10 = **17**

34) 8 + **5** = 13

35) 18 − 12 = **6**

36) 19 − 6 = **13**

37) **19** + 0 = 19

38) 18 − 2 = **16**

39) 7 − **1** = 6

40) 1 + 13 = **14**

41) **14** − 2 = 12

42) 15 + 4 = **19**

43) **2** + 15 = 17

44) 15 − 15 = **0**

45) **3** + 16 = 19

46) **18** − 8 = 10

47) 10 + 4 = **14**

48) **16** − 5 = 11

49) 20 − **6** = 14

50) 10 + 10 = **20**

51) 20 + **0** = 20

52) 11 − **9** = 2

53) 17 − **6** = 11

54) 15 − **5** = 10

55) 2 + **2** = 4

56) 5 + 13 = **18**

57) **3** − 0 = 3

58) 17 + **1** = 18

59) 6 − **4** = 2

60) 15 + **4** = 19

Date : _________   Name : _________

Time : _________   Score : _____/60

Mixed Problems 0 - 20

1) 1 + 19 = _____   2) 5 − 5 = _____   3) 10 + 6 = _____

4) 6 − 1 = _____   5) 20 − 19 = _____   6) 1 + 4 = _____

7) 18 − 1 = _____   8) 4 + 15 = _____   9) 6 + 12 = _____

10) 20 + 0 = _____   11) 19 − 16 = _____   12) 8 − 8 = _____

13) 8 + 4 = _____   14) 20 + 0 = _____   15) 13 − 6 = _____

16) 14 − 7 = _____   17) 17 − 8 = _____   18) 4 − 1 = _____

19) 7 + 8 = _____   20) 7 + 13 = _____   21) 7 − 6 = _____

22) 16 − 16 = _____   23) 13 + 4 = _____   24) 17 + 3 = _____

25) 15 + 2 = _____   26) 12 − 8 = _____   27) 8 − 1 = _____

28) 1 + 6 = _____   29) 5 + 3 = _____   30) 1 + 3 = _____

31) 19 − 6 = _____   32) 16 − 6 = _____   33) 15 − 3 = _____

34) 0 + 20 = _____   35) 11 − 7 = _____   36) 5 + 12 = _____

37) 7 − 5 = _____   38) 16 + 2 = _____   39) 10 + 9 = _____

40) 2 − 0 = _____   41) 20 − 10 = _____   42) 0 + 19 = _____

43) 6 + 12 = _____   44) 14 − 12 = _____   45) 1 + 17 = _____

46) 4 − 2 = _____   47) 6 − 6 = _____   48) 3 + 16 = _____

49) 6 + 12 = _____   50) 13 − 13 = _____   51) 14 − 14 = _____

52) 14 + 4 = _____   53) 20 − 9 = _____   54) 7 + 4 = _____

55) 12 − 4 = _____   56) 6 + 5 = _____   57) 20 + 0 = _____

58) 18 − 8 = _____   59) 10 + 8 = _____   60) 6 − 5 = _____

Answer Key
Mixed Problems 0 - 20

1) $1 + 19 = \textbf{20}$

2) $5 - 5 = \textbf{0}$

3) $10 + 6 = \textbf{16}$

4) $6 - 1 = \textbf{5}$

5) $20 - 19 = \textbf{1}$

6) $1 + 4 = \textbf{5}$

7) $18 - 1 = \textbf{17}$

8) $4 + 15 = \textbf{19}$

9) $6 + 12 = \textbf{18}$

10) $20 + 0 = \textbf{20}$

11) $19 - 16 = \textbf{3}$

12) $8 - 8 = \textbf{0}$

13) $8 + 4 = \textbf{12}$

14) $20 + 0 = \textbf{20}$

15) $13 - 6 = \textbf{7}$

16) $14 - 7 = \textbf{7}$

17) $17 - 8 = \textbf{9}$

18) $4 - 1 = \textbf{3}$

19) $7 + 8 = \textbf{15}$

20) $7 + 13 = \textbf{20}$

21) $7 - 6 = \textbf{1}$

22) $16 - 16 = \textbf{0}$

23) $13 + 4 = \textbf{17}$

24) $17 + 3 = \textbf{20}$

25) $15 + 2 = \textbf{17}$

26) $12 - 8 = \textbf{4}$

27) $8 - 1 = \textbf{7}$

28) $1 + 6 = \textbf{7}$

29) $5 + 3 = \textbf{8}$

30) $1 + 3 = \textbf{4}$

31) $19 - 6 = \textbf{13}$

32) $16 - 6 = \textbf{10}$

33) $15 - 3 = \textbf{12}$

34) $0 + 20 = \textbf{20}$

35) $11 - 7 = \textbf{4}$

36) $5 + 12 = \textbf{17}$

37) $7 - 5 = \textbf{2}$

38) $16 + 2 = \textbf{18}$

39) $10 + 9 = \textbf{19}$

40) $2 - 0 = \textbf{2}$

41) $20 - 10 = \textbf{10}$

42) $0 + 19 = \textbf{19}$

43) $6 + 12 = \textbf{18}$

44) $14 - 12 = \textbf{2}$

45) $1 + 17 = \textbf{18}$

46) $4 - 2 = \textbf{2}$

47) $6 - 6 = \textbf{0}$

48) $3 + 16 = \textbf{19}$

49) $6 + 12 = \textbf{18}$

50) $13 - 13 = \textbf{0}$

51) $14 - 14 = \textbf{0}$

52) $14 + 4 = \textbf{18}$

53) $20 - 9 = \textbf{11}$

54) $7 + 4 = \textbf{11}$

55) $12 - 4 = \textbf{8}$

56) $6 + 5 = \textbf{11}$

57) $20 + 0 = \textbf{20}$

58) $18 - 8 = \textbf{10}$

59) $10 + 8 = \textbf{18}$

60) $6 - 5 = \textbf{1}$

Mixed Problems 0 - 20

1) $17 + \underline{\hspace{1cm}} = 19$

2) $\underline{\hspace{1cm}} - 10 = 6$

3) $\underline{\hspace{1cm}} + 12 = 14$

4) $17 - \underline{\hspace{1cm}} = 11$

5) $2 + \underline{\hspace{1cm}} = 3$

6) $\underline{\hspace{1cm}} - 2 = 16$

7) $9 + \underline{\hspace{1cm}} = 19$

8) $\underline{\hspace{1cm}} - 10 = 7$

9) $\underline{\hspace{1cm}} + 17 = 19$

10) $16 - 10 = \underline{\hspace{1cm}}$

11) $\underline{\hspace{1cm}} - 0 = 1$

12) $2 + \underline{\hspace{1cm}} = 9$

13) $\underline{\hspace{1cm}} - 2 = 10$

14) $\underline{\hspace{1cm}} + 17 = 20$

15) $19 - 4 = \underline{\hspace{1cm}}$

16) $\underline{\hspace{1cm}} + 1 = 1$

17) $6 + 12 = \underline{\hspace{1cm}}$

18) $\underline{\hspace{1cm}} + 2 = 8$

19) $\underline{\hspace{1cm}} - 14 = 3$

20) $14 - 3 = \underline{\hspace{1cm}}$

21) $\underline{\hspace{1cm}} - 8 = 5$

22) $8 + 8 = \underline{\hspace{1cm}}$

23) $8 + 5 = \underline{\hspace{1cm}}$

24) $\underline{\hspace{1cm}} - 3 = 3$

25) $\underline{\hspace{1cm}} - 1 = 17$

26) $18 + \underline{\hspace{1cm}} = 18$

27) $10 - 3 = \underline{\hspace{1cm}}$

28) $\underline{\hspace{1cm}} + 6 = 10$

29) $12 - 7 = \underline{\hspace{1cm}}$

30) $\underline{\hspace{1cm}} + 8 = 18$

31) $6 + \underline{\hspace{1cm}} = 14$

32) $\underline{\hspace{1cm}} - 10 = 7$

33) $6 + 3 = \underline{\hspace{1cm}}$

34) $\underline{\hspace{1cm}} - 6 = 6$

35) $20 + \underline{\hspace{1cm}} = 20$

36) $\underline{\hspace{1cm}} - 6 = 3$

37) $16 - \underline{\hspace{1cm}} = 16$

38) $6 + \underline{\hspace{1cm}} = 7$

39) $0 + \underline{\hspace{1cm}} = 0$

40) $9 - 4 = \underline{\hspace{1cm}}$

41) $\underline{\hspace{1cm}} + 11 = 16$

42) $\underline{\hspace{1cm}} - 8 = 1$

43) $\underline{\hspace{1cm}} - 1 = 11$

44) $\underline{\hspace{1cm}} + 7 = 8$

45) $\underline{\hspace{1cm}} + 4 = 20$

46) $\underline{\hspace{1cm}} - 1 = 4$

47) $12 - 12 = \underline{\hspace{1cm}}$

48) $5 + 12 = \underline{\hspace{1cm}}$

49) $\underline{\hspace{1cm}} - 5 = 7$

50) $\underline{\hspace{1cm}} + 1 = 5$

51) $18 - 4 = \underline{\hspace{1cm}}$

52) $1 + \underline{\hspace{1cm}} = 20$

53) $\underline{\hspace{1cm}} + 1 = 15$

54) $\underline{\hspace{1cm}} - 9 = 4$

55) $17 + \underline{\hspace{1cm}} = 20$

56) $9 - \underline{\hspace{1cm}} = 9$

57) $5 + 7 = \underline{\hspace{1cm}}$

58) $15 + 3 = \underline{\hspace{1cm}}$

59) $\underline{\hspace{1cm}} - 1 = 7$

60) $\underline{\hspace{1cm}} - 6 = 4$

Answer Key
Mixed Problems 0 - 20

1) $17 + 2 = 19$

2) $16 - 10 = 6$

3) $2 + 12 = 14$

4) $17 - 6 = 11$

5) $2 + 1 = 3$

6) $18 - 2 = 16$

7) $9 + 10 = 19$

8) $17 - 10 = 7$

9) $2 + 17 = 19$

10) $16 - 10 = 6$

11) $1 - 0 = 1$

12) $2 + 7 = 9$

13) $12 - 2 = 10$

14) $3 + 17 = 20$

15) $19 - 4 = 15$

16) $0 + 1 = 1$

17) $6 + 12 = 18$

18) $6 + 2 = 8$

19) $17 - 14 = 3$

20) $14 - 3 = 11$

21) $13 - 8 = 5$

22) $8 + 8 = 16$

23) $8 + 5 = 13$

24) $6 - 3 = 3$

25) $18 - 1 = 17$

26) $18 + 0 = 18$

27) $10 - 3 = 7$

28) $4 + 6 = 10$

29) $12 - 7 = 5$

30) $10 + 8 = 18$

31) $6 + 8 = 14$

32) $17 - 10 = 7$

33) $6 + 3 = 9$

34) $12 - 6 = 6$

35) $20 + 0 = 20$

36) $9 - 6 = 3$

37) $16 - 0 = 16$

38) $6 + 1 = 7$

39) $0 + 0 = 0$

40) $9 - 4 = 5$

41) $5 + 11 = 16$

42) $9 - 8 = 1$

43) $12 - 1 = 11$

44) $1 + 7 = 8$

45) $16 + 4 = 20$

46) $5 - 1 = 4$

47) $12 - 12 = 0$

48) $5 + 12 = 17$

49) $12 - 5 = 7$

50) $4 + 1 = 5$

51) $18 - 4 = 14$

52) $1 + 19 = 20$

53) $14 + 1 = 15$

54) $13 - 9 = 4$

55) $17 + 3 = 20$

56) $9 - 0 = 9$

57) $5 + 7 = 12$

58) $15 + 3 = 18$

59) $8 - 1 = 7$

60) $10 - 6 = 4$

Mixed Problems 0 - 20

1) $3 + \_\_\_\_\_ = 12$

2) $\_\_\_\_\_ - 3 = 13$

3) $3 + \_\_\_\_\_ = 10$

4) $11 - 2 = \_\_\_\_\_$

5) $20 - \_\_\_\_\_ = 20$

6) $\_\_\_\_\_ + 8 = 17$

7) $19 - \_\_\_\_\_ = 17$

8) $13 + \_\_\_\_\_ = 20$

9) $13 + 6 = \_\_\_\_\_$

10) $16 - \_\_\_\_\_ = 4$

11) $1 + \_\_\_\_\_ = 17$

12) $\_\_\_\_\_ - 7 = 4$

13) $3 + 14 = \_\_\_\_\_$

14) $\_\_\_\_\_ + 3 = 15$

15) $11 - 4 = \_\_\_\_\_$

16) $15 - \_\_\_\_\_ = 13$

17) $17 - \_\_\_\_\_ = 12$

18) $\_\_\_\_\_ + 8 = 12$

19) $\_\_\_\_\_ + 7 = 20$

20) $\_\_\_\_\_ - 13 = 4$

21) $2 + 6 = \_\_\_\_\_$

22) $6 - 1 = \_\_\_\_\_$

23) $13 - \_\_\_\_\_ = 12$

24) $11 + \_\_\_\_\_ = 14$

25) $\_\_\_\_\_ + 1 = 18$

26) $19 - 6 = \_\_\_\_\_$

27) $12 - \_\_\_\_\_ = 6$

28) $3 + 16 = \_\_\_\_\_$

29) $3 + \_\_\_\_\_ = 19$

30) $16 - \_\_\_\_\_ = 3$

31) $9 + \_\_\_\_\_ = 15$

32) $20 - \_\_\_\_\_ = 5$

33) $15 - \_\_\_\_\_ = 1$

34) $20 - 1 = \_\_\_\_\_$

35) $\_\_\_\_\_ + 14 = 17$

36) $\_\_\_\_\_ + 3 = 16$

37) $5 + \_\_\_\_\_ = 6$

38) $\_\_\_\_\_ + 8 = 17$

39) $13 - \_\_\_\_\_ = 1$

40) $\_\_\_\_\_ - 10 = 1$

41) $\_\_\_\_\_ + 12 = 16$

42) $14 - 7 = \_\_\_\_\_$

43) $11 + 5 = \_\_\_\_\_$

44) $11 - \_\_\_\_\_ = 8$

45) $16 - \_\_\_\_\_ = 9$

46) $\_\_\_\_\_ + 11 = 17$

47) $\_\_\_\_\_ - 4 = 14$

48) $15 + \_\_\_\_\_ = 17$

49) $9 + \_\_\_\_\_ = 12$

50) $1 + \_\_\_\_\_ = 18$

51) $12 - \_\_\_\_\_ = 2$

52) $12 - \_\_\_\_\_ = 6$

53) $18 - \_\_\_\_\_ = 6$

54) $20 - 18 = \_\_\_\_\_$

55) $5 + \_\_\_\_\_ = 18$

56) $11 + 1 = \_\_\_\_\_$

57) $17 - 16 = \_\_\_\_\_$

58) $\_\_\_\_\_ + 9 = 19$

59) $15 - 13 = \_\_\_\_\_$

60) $16 + \_\_\_\_\_ = 16$

Answer Key

Mixed Problems 0 - 20

1) $3 + \mathbf{9} = 12$

2) $\mathbf{16} - 3 = 13$

3) $3 + \mathbf{7} = 10$

4) $11 - 2 = \mathbf{9}$

5) $20 - \mathbf{0} = 20$

6) $\mathbf{9} + 8 = 17$

7) $19 - \mathbf{2} = 17$

8) $13 + \mathbf{7} = 20$

9) $13 + 6 = \mathbf{19}$

10) $16 - \mathbf{12} = 4$

11) $1 + \mathbf{16} = 17$

12) $\mathbf{11} - 7 = 4$

13) $3 + 14 = \mathbf{17}$

14) $\mathbf{12} + 3 = 15$

15) $11 - 4 = \mathbf{7}$

16) $15 - \mathbf{2} = 13$

17) $17 - \mathbf{5} = 12$

18) $\mathbf{4} + 8 = 12$

19) $\mathbf{13} + 7 = 20$

20) $\mathbf{17} - 13 = 4$

21) $2 + 6 = \mathbf{8}$

22) $6 - 1 = \mathbf{5}$

23) $13 - \mathbf{1} = 12$

24) $11 + \mathbf{3} = 14$

25) $\mathbf{17} + 1 = 18$

26) $19 - 6 = \mathbf{13}$

27) $12 - \mathbf{6} = 6$

28) $3 + 16 = \mathbf{19}$

29) $3 + \mathbf{16} = 19$

30) $16 - \mathbf{13} = 3$

31) $9 + \mathbf{6} = 15$

32) $20 - \mathbf{15} = 5$

33) $15 - \mathbf{14} = 1$

34) $20 - 1 = \mathbf{19}$

35) $\mathbf{3} + 14 = 17$

36) $\mathbf{13} + 3 = 16$

37) $5 + \mathbf{1} = 6$

38) $\mathbf{9} + 8 = 17$

39) $13 - \mathbf{12} = 1$

40) $\mathbf{11} - 10 = 1$

41) $\mathbf{4} + 12 = 16$

42) $14 - 7 = \mathbf{7}$

43) $11 + 5 = \mathbf{16}$

44) $11 - \mathbf{3} = 8$

45) $16 - \mathbf{7} = 9$

46) $\mathbf{6} + 11 = 17$

47) $\mathbf{18} - 4 = 14$

48) $15 + \mathbf{2} = 17$

49) $9 + \mathbf{3} = 12$

50) $1 + \mathbf{17} = 18$

51) $12 - \mathbf{10} = 2$

52) $12 - \mathbf{6} = 6$

53) $18 - \mathbf{12} = 6$

54) $20 - 18 = \mathbf{2}$

55) $5 + \mathbf{13} = 18$

56) $11 + 1 = \mathbf{12}$

57) $17 - 16 = \mathbf{1}$

58) $\mathbf{10} + 9 = 19$

59) $15 - 13 = \mathbf{2}$

60) $16 + \mathbf{0} = 16$

Mixed Problems 0 - 20

1) 16 − 5 = _____	2) 15 + 1 = _____	3) 6 − 1 = _____
4) 17 + 3 = _____	5) 7 − 2 = _____	6) 8 + 5 = _____
7) 8 + 6 = _____	8) 18 − 1 = _____	9) 13 + 7 = _____
10) 16 − 5 = _____	11) 10 − 6 = _____	12) 5 + 4 = _____
13) 14 − 8 = _____	14) 16 + 1 = _____	15) 17 + 2 = _____
16) 15 − 14 = _____	17) 19 + 1 = _____	18) 17 − 1 = _____
19) 18 + 2 = _____	20) 14 − 12 = _____	21) 20 − 6 = _____
22) 2 + 4 = _____	23) 7 − 6 = _____	24) 9 + 9 = _____
25) 4 + 15 = _____	26) 20 − 11 = _____	27) 5 − 1 = _____
28) 1 + 8 = _____	29) 13 + 5 = _____	30) 14 − 11 = _____
31) 15 + 3 = _____	32) 18 − 15 = _____	33) 18 − 18 = _____
34) 14 − 13 = _____	35) 1 + 4 = _____	36) 8 + 12 = _____
37) 2 + 3 = _____	38) 7 + 12 = _____	39) 19 − 10 = _____
40) 11 − 2 = _____	41) 11 + 4 = _____	42) 5 + 9 = _____
43) 15 − 12 = _____	44) 18 − 12 = _____	45) 10 − 3 = _____
46) 11 − 5 = _____	47) 1 + 11 = _____	48) 4 + 12 = _____
49) 15 + 4 = _____	50) 9 + 11 = _____	51) 19 − 11 = _____
52) 9 − 6 = _____	53) 8 + 9 = _____	54) 4 − 4 = _____
55) 1 + 18 = _____	56) 10 − 8 = _____	57) 5 − 1 = _____
58) 5 + 16 = _____	59) 19 − 12 = _____	60) 3 + 14 = _____

Answer Key
Mixed Problems 0 - 20

1) 16 − 5 = **11**

2) 15 + 1 = **16**

3) 6 − 1 = **5**

4) 17 + 3 = **20**

5) 7 − 2 = **5**

6) 8 + 5 = **13**

7) 8 + 6 = **14**

8) 18 − 1 = **17**

9) 13 + 7 = **20**

10) 16 − 5 = **11**

11) 10 − 6 = **4**

12) 5 + 4 = **9**

13) 14 − 8 = **6**

14) 16 + 1 = **17**

15) 17 + 2 = **19**

16) 15 − 14 = **1**

17) 19 + 1 = **20**

18) 17 − 1 = **16**

19) 18 + 2 = **20**

20) 14 − 12 = **2**

21) 20 − 6 = **14**

22) 2 + 4 = **6**

23) 7 − 6 = **1**

24) 9 + 9 = **18**

25) 4 + 15 = **19**

26) 20 − 11 = **9**

27) 5 − 1 = **4**

28) 1 + 8 = **9**

29) 13 + 5 = **18**

30) 14 − 11 = **3**

31) 15 + 3 = **18**

32) 18 − 15 = **3**

33) 18 − 18 = **0**

34) 14 − 13 = **1**

35) 1 + 4 = **5**

36) 8 + 12 = **20**

37) 2 + 3 = **5**

38) 7 + 12 = **19**

39) 19 − 10 = **9**

40) 11 − 2 = **9**

41) 11 + 4 = **15**

42) 5 + 9 = **14**

43) 15 − 12 = **3**

44) 18 − 12 = **6**

45) 10 − 3 = **7**

46) 11 − 5 = **6**

47) 1 + 11 = **12**

48) 4 + 12 = **16**

49) 15 + 4 = **19**

50) 9 + 11 = **20**

51) 19 − 11 = **8**

52) 9 − 6 = **3**

53) 8 + 9 = **17**

54) 4 − 4 = **0**

55) 1 + 18 = **19**

56) 10 − 8 = **2**

57) 5 − 1 = **4**

58) 5 + 16 = **21**

59) 19 − 12 = **7**

60) 3 + 14 = **17**